DESIGN CHALLENGES

Written by

Michelle Powers, Teri Barenborg,
Tari Sexton, and Lauren Monroe

Editors: Christie Weltz and Jasmine Tabrizi
Designer/Production: Kammy Peyton
Art Director: Moonhee Pak
Project Director: Stacey Faulkner

DEDICATION

This book is dedicated to all of the educators and children who have inspired us to make education a hands-on experience and, most importantly, instilled within us a lifelong love of learning.

ACKNOWLEDGMENTS

First and foremost, we would like to thank our families and friends who have supported us in so many ways—from the steadfast support of our chosen career path and passion all the way through the inspiration and creation of this series of books. Each of us has an amazing support system that has not only encouraged us but also made it possible for us to devote our time to this project. A sincere thank-you to our colleagues, both past and present, as well as all the educators who have inspired us to create a collection of lessons that encourage students to grow and take ownership of their learning. Without the continued support and encouragement of our dear friend Lynn Howard, these books would not have been possible.

Our school district, St. Lucie Public Schools, known for being the first Kids at Hope school district in the state of Florida, motivated us to build a culture of learning where students state daily that "All children are capable of success. No exceptions." This mindset, along with the work of Carol Dweck and her focus on self-efficacy through a growth mindset, has inspired us to develop lessons that encourage problem solving and perseverance, allowing students to learn from their mistakes.

We would like to thank the various teachers who have opened their doors to us and, more importantly, the students in those classrooms who have tested these exciting lessons during their development. These teachers have allowed us to model, motivate, and encourage them to transition from the "Sage on the Stage" to a "Guide on the Side," giving students the opportunity to drive their own learning.

FOREWORD

Science instruction has changed. Many of us can remember the traditional lecture and note giving model of instruction that had been used for years. I was very alone in my middle school earth science classroom and had no support, no textbook or curriculum guide. Living day to day with content that was totally unfamiliar to me, I taught the same way to all students and didn't realize that many of them were not engaged or learning. I had to change and allow for more engagement, exploration, and experimentation. It quickly became the way I taught, and students benefited from the problem solving, collaboration, and inquiry-based activities. When I began my science teaching career years ago, I would have appreciated a resource that provided me with a set of classroom lessons that would challenge and motivate my students.

The Next Generation Science Standards are placing a great emphasis on how we "do science" in the classroom. The integration of science, technology, engineering, arts, and math (STEAM) provides multiple opportunities to include problem solving, engineering practices, and literacy while engaging and motivating students in real-world science experiences.

I really like this book. These lessons are perfect for any teacher who may or may not feel comfortable with teaching science. I really like that the lessons are aligned with the 5E Instructional Model (engage, explore, explain, elaborate, and evaluate). Teachers who use the lessons will address the 5E model and challenge their students with the engineering process. The authors are a team of educators who understand how to teach science. Their teaching has evolved from a traditional approach to becoming facilitators of science knowledge. Teri, Lauren, Michelle, and Tari have spent time learning about the changes in science education and how to design effective science classroom environments. As a professional development associate, I spent three years with them as they explored how to create a balanced science program focused on the Next Generation Science Standards. They invested a large amount of time researching what works and implementing those best practices in their classrooms. I have had the opportunity to be in all of their classrooms and see the engagement and excitement as students collaborate on real-world engineering design problems. The teachers continually reinforce the idea that their students ARE scientists and must practice the habits of scientists. A by-product of these teachers' efforts is a book that other teachers can use today in their classrooms to make it exciting to teach and learn about science!

I am honored that Teri, Lauren, Michelle, and Tari asked me to write the foreword for their book. These teachers truly live and breathe quality science teaching and learning. Their passion, dedication, and commitment to effective science instruction make the activities and ideas in this book invaluable to anyone who wants to get excited about STEAM in their classroom.

Lynn F. Howard
Author and Professional Development Associate
Five Easy Steps to a Balanced Science Program

TABLE OF CONTENTS

GETTING STARTED

Introduction 5
How to Use This Book 6
Standards 13
Integration in the Engineering
Design Challenge 14
STEAM Design Process 15
Recording Information
in a Science Notebook 16

LIFE SCIENCE

Last Critter Standing 74
Extraterrestrial Gardening 80
Living Cells 86
Live on the Scene 92

EARTH AND SPACE SCIENCE

A World for the Lorax 20
How Far Must I Go? 26
Weather Detector 32
Shifting Mines 38

PHYSICAL SCIENCE

Magically Move It! 98
Modify the Room 104
Land Yachts Ahoy! 110
Smooth as Ice 116
Zipping Along 122

ENGINEERING DESIGN

Who Done It? 44
Engineer That! 50
Riveting Roller Coasters 56
Shake It Up! 62
We Built a Zoo 68

APPENDIX

Lesson Plan-Specific Reproducibles 129
Individual Blueprint Design Sheet 143
Group Blueprint Design Sheet 144
Budget Planning Chart 145
STEAM Job Cards 146
STEAM Rubric 147
Bibliography 149

INTRODUCTION

Science, technology, engineering, art, and math work together to make learning fun!

The Next Generation Science Standards place a greater emphasis on science, technology, engineering, arts, and math (STEAM) in today's classrooms. Schools are implementing and encouraging strong STEAM programs in classrooms in order to provide critical thinking lessons that meet the content standards. STEAM lessons should include problem-solving skills, enhance learning across various disciplines, promote student inquiry, and engage students with real-world situations. Students should be exposed to careers in the STEAM fields and develop skills such as communication, data analysis, following a process, designing a product, and argumentation based on evidence, all while cementing effective collaboration techniques that are necessary for a successful career in STEAM fields.

The lessons in this book are intended to support teachers in implementing the engineering design process in their classroom while integrating national standards from other disciplines. In the engineering design process, teachers become a facilitator rather than the instructional focus. Teachers encourage and guide students to work as a team to find a creative solution without providing step-by-step instructions. The engineering design process shifts away from the long-standing process of the scientific method by placing more emphasis on inquiry. Students are inspired to act as scientists and engineers through the use of sketches, diagrams, mathematical relationships, and literacy connections. By creating their very own models and products based on background information from their studies, students are immediately engaged through a meaningful, rewarding lesson.

Each lesson begins by presenting students with a design challenge scenario, or hook, in order to immediately excite students with a real-world situation that they are on a mission to solve. Students are then given a dilemma, mission, and blueprint design sheet and are asked to collaborate with team members to create several prototypes. Teams are required to choose one prototype to present to their teacher before gathering materials and constructing the chosen design. After testing out their design, teams take part in a class discussion and modify their ideas for redesign and improvement of their prototype. Finally, teams are asked to create a justification piece in order to sell their new prototype. Suggestions for justification projects are provided for each design challenge and include writing a persuasive letter, creating an advertisement or presentation, recording a video, or any other creative ideas they come up with in response to the challenge.

The engaging STEAM design challenge lessons in this book

- Promote analytical and reflective thinking
- Enhance learning across various disciplines
- Encourage students to collaborate to solve real-world design challenges
- Integrate national standards
- Are classroom tested

HOW TO USE THIS BOOK

STEAM design challenges follow the engineering practices that have become recently known in the education field. Engineering practices teach students to solve a problem by designing, creating, and justifying their design. With this model in mind, teachers shift from a "giver of information" to a "facilitator of knowledge." Instead of leading children to the right conclusion through experimental steps, the teacher allows them to work through the process themselves, often changing their plan to improve their original design.

STEAM design challenges allow art to support and enhance the learning of science and math while the engineering process is followed. Students will often use, or be encouraged to use, technology to facilitate their learning. The teacher's role as facilitator allows him or her to guide student thinking by asking questions instead of giving answers. Each lesson covers cross-curricular standards and supports teacher planning for collaboration with other teachers.

Typically, science is not taught as often in elementary school as English, reading, writing, and math, so assignments have been included within the lessons that will assist in giving students skills and practice in those other key subjects.

Lessons focus on key national science standards that are required for many standardized tests and include core English language arts and math standards. National engineering standards as well as national arts and national technology standards are also included in the lessons.

The 5E Instructional Model emphasizes building new ideas using existing knowledge. The components of this model—*Engage, Explore, Explain, Elaborate,* and *Evaluate*—are also a key design feature in the structure of each design challenge. Each design challenge requires the students to respond using mathematical, written, oral, and theatrical skills that are developmentally appropriate while working through each phase of the 5E model.

PHASES OF THE 5E MODEL

ENGAGE
Students make connections between past and present learning and focus their thinking on learning outcomes in the activity.

EXPLORE
Students continue to build on their knowledge of their learning through exploration and manipulation of materials.

EXPLAIN
Students support their understanding of the concepts through verbal or written communication. This is also a time when students may demonstrate new skills and when teachers can introduce new vocabulary.

ELABORATE
Students extend their understanding of concepts by obtaining more information about a topic through new experiences.

EVALUATE
Students assess their understanding of key concepts and skills.

LESSON PLAN FORMAT

Each lesson centers around the Design Challenge Purpose and has two distinct sections—Setting the Stage and STEAM in Action.

- **Setting the Stage** provides an overview of the lesson, suggested time frame, the background knowledge needed for the teacher and students as well as the standards, target vocabulary, and materials needed.

- **STEAM in Action** outlines the step-by-step procedure for implementing the lesson.

LESSON PLAN COMPONENTS

SETTING THE STAGE

Header: This section includes the title, suggested time frame for completing the lesson, and the STEAM acronym, in which the capital letters denote the main disciplines that are highlighted in each particular lesson.

Time: A suggested approximate total time for completing each lesson is provided. Because the amount of time teachers have to teach science varies within different states, districts, schools, and even grade levels, you may need to break up the lesson into smaller segments over the course of several days. Natural breaks occur between design and construction, between construction and testing, and between testing and justification.

You may choose to use the lesson ideas in the Student Development section to deepen prior knowledge, or you may have your students use the literacy connections and any reputable websites you are familiar with. The lesson ideas in the Justification section are included as an optional extension of the core lesson. None of the activities before or after the core lesson are included in the time estimates. Refer to the suggested lesson timeline on page 11.

Design Challenge Purpose: This is the statement that sets the stage for the design challenge and outlines student objectives and expectations for what they should learn by completing the design challenge.

Teacher Development: This section provides background information about the science content being addressed in the lesson. Information included assists the teacher in understanding key science concepts. We understand that professional development at the elementary teacher level is often geared toward instructional delivery instead of content, especially in the content area of science. This section is provided to help support teachers who may not be as familiar with science content.

Student Development: This section contains a description of the concepts students will need to understand to complete the design challenge successfully. A link to the STEAM Dreamers website, which has active web links and additional suggested lesson ideas for deepening students' understanding of relevant science concepts, can be found on the inside front cover of this book.

Standards: This section lists specific standards for science, technology, engineering, art, math, and English language arts, along with the science and engineering practices and crosscutting concepts. These standards may apply to the activities in the challenges or in the justifications that follow. Please make sure that you review the standards for each of the lessons. The website for each set of standards is listed on page 13.

Target Vocabulary: This section lists target vocabulary to support and enhance the lesson content and to deepen students' understanding of the terms. These vocabulary terms are related to the academic content that the design challenge focuses on; can be used throughout the design challenge when in group discussion; and are an integral component of the standards covered in the challenge. Reviewing the target vocabulary prior to beginning the design challenge is recommended as students need to apply their knowledge of the science concepts and target vocabulary when solving the challenges. Ultimately, the target vocabulary should be revisited multiple times throughout the lesson.

Materials: This section lists materials and equipment that have been selected for the lessons. All materials are meant to be easy to find, inexpensive to purchase, recycled, or commonly available for free. Substitute with similar items if you have them on hand, or visit www.SteamDreamers.com for substitute suggestions.

Literacy Connections: This section lists books or articles that are meant to be used with students prior to the design challenge in order to strengthen their background knowledge and to enhance the integration of literacy in STEAM. These connections can be used during the daily classroom reading block or during small- and/or whole-group instruction.

Current literacy connections for each lesson can be accessed through our website: www.SteamDreamers.com.

STEAM IN ACTION

The Dilemma: This section includes a unique real-world dilemma or scenario that hooks the students and gets them excited to solve the problem. The dilemma may include a plausible circumstance or a wild story designed to make them think. When planning the design of their prototype, student should ask themselves questions such as *Who is the client? What do we need to create? What is the purpose of the creation? What is the ultimate goal?* Students should discuss these questions with other members of their team and record their responses in their science notebooks.

Note: This is the Engage portion of the lesson, as outlined in the 5E Instructional Model.

The Mission: This section includes the defined challenge statement. This is ultimately the goal that the students are trying to reach.

Blueprint Design: This section instructs students on how to focus their thinking in order to solve the problem. Individual team members design their own plans for prototypes and list the pros and cons of their designs. Each team member reviews the Individual Blueprint Design Sheet of every other team member and records the pros and cons he or she sees. The team then chooses which member's design it will move forward with. This is where students have the opportunity to discuss and make decisions based on their analysis on the Individual Blueprint Design Sheets. Students are allowed and encouraged to add their artistic touches to their thinking. Individual and Group Blueprint Design Sheets are found in the Appendix.

Note: This is the Explore portion of the lesson, as outlined in the 5E Instructional Model.

Engineering Design Process: In this section of the lesson, teams will take their group's selected prototype through the engineering design process to create, test, analyze, and redesign as necessary until they have successfully completed their mission.

- The first step in the process is the Engineering Task in which teams will engineer their prototype.

- Students will then test their prototype based upon the mission statement.

- The analysis of their testing will include data collection and determination of success.

- The Redesign and Retest cycle will continue until the team has successfully completed the mission.

Helpful Tips: In this section you'll find suggestions designed to address common issues that may arise during the design challenges. Some tips are geared toward the steps in the engineering design process, and some are more lesson-specific.

Reflections: This section provides suggestions for reflective questions to ask students to help guide and facilitate their thinking at various stages within the engineering design process. It is recommended that students record these questions and their reflections in a science notebook. See pages 16–19 for more information on using a science notebook.

Note: This is the Explain and Elaborate portion of the lesson, as outlined in the 5E Instructional Model.

Justification: This is the stage of the lesson where students apply what they learned in a meaningful and creative way through different mediums, such as technology and the arts. These justifications can occur in many forms: a formal letter, an advertisement, a poem, a jingle, a skit, or a technology-enhanced presentation.

Note: This is the Evaluate portion of the lesson, as outlined in the 5E Instructional Model.

SUGGESTED LESSON TIMELINE

Lesson Progression:

1. Teacher Development/Student Development/Literacy Connections

2. Dilemma/Mission/Blueprint Design

3. Engineering Task/Test Trial/Analyze/ Redesign/Reflection

4. Justification

If the lesson will be spread out over multiple days:

Day 1: Teacher Development/Student Development/Literacy Connections

Day 2: Dilemma/Mission/Blueprint Design

Day 3: Engineering Task/Test Trial

Days 4-6: Analyze/Redesign/Reflection (Can be spread over 3 days)

Days 7-8: Justification

THE APPENDIX

Lesson-Specific Activity Pages:
Some lessons include specific activity pages for enhancing or completing the design challenges. They are found in the Appendix section.

Blueprint Design Sheets:
Every lesson requires students to first use the Individual Blueprint Design Sheet to create and list the pros and cons of their and their teammates' designs. Students will discuss their designs with team members and choose one design to use for building their prototype. This design, and reasons why it was chosen, are recorded on the Group Blueprint Design Sheet.

Budget Planning Chart:
Any of the lessons can implement a budget for an added mathematical challenge. Prior to the start of the challenge, assign each material a cost and display the costs for the class to reference throughout the challenge. Then decide on an overall budget for the materials. Students can use the Budget Planning Chart to itemize materials and identify the total cost of the materials needed to complete the

challenge. The chart is blank to allow for more flexibility with the materials needed for specific challenges. Ensure students have a limit to what they can spend during the challenge. You can chose not to incorporate a budget if you are short on time. The time needed to assign specific material costs is not included in the overall completion time for the lessons.

Rubric: A rubric for grading the STEAM challenges is included. This rubric focuses on the engineering process. However, it does not include a means to assess the justification components.

	EXEMPLARY	PROFICIENT	PROGRESSING	BEGINNING
STEAM DESIGN CHALLENGES TEAM RUBRIC				
DESIGN	Team members reach consensus as to which prototype to construct. They complete team blueprint design sheet in which they include their reasons for selecting the team prototype. They include a written explanation to compare and contrast the prototypes they sketched individually. Prototype is constructed according to specifications in the team blueprint design.	Team members reach consensus as to which prototype to construct. They include their reasons for selecting the prototype but do not include a written explanation to compare and contrast the prototypes they sketched individually. Prototype is constructed according to the specifications in the team blueprint design.	Team members reach consensus as to which prototype to construct. They include their reasons for selecting the prototype but do not include a written explanation to compare and contrast the prototypes they sketched individually. Prototype is not constructed according to the specifications of the blueprint design.	Team members reach consensus as to which prototype to construct. They do not include either their reasons for selecting the prototype or a written explanation to compare and contrast the prototypes they sketched. Prototype is constructed.
TEST	Team tests its prototype. Team members record observations that align with the design challenge. They make note of any unique design flaws.	Team tests its prototype and records observations that align with the design challenge.	Team tests its prototype. Team members record observations that do not align with the design challenge.	Team tests its prototype. Team members do not record observations.

STEAM Job Cards: If your students are struggling with the collaboration process, try assigning them specific roles. Suggestions for jobs are provided on the STEAM Job Cards. Four students per team is recommended. The Accounts Manager role will only occur during the design challenges that involve a budget. In these cases, one student will have two roles, one of which is the Accounts Manager.

STEAM Money: The use of STEAM money is a fun way to engage students and connect the design challenges that incorporate budgets to the real world by having teams "purchase" materials. The use of STEAM money is completely optional. The following suggestions are offered should you choose to incorporate STEAM money into any of your lessons.

- Design and create your own STEAM money.

- Print multiple copies and laminate for durability and multiple use.

- Enlist the help of a parent volunteer to prepare the STEAM money at the beginning of the year.

- Assign material costs with the class before beginning lessons with budgets, or incorporate this into your long-range planning before school begins. This only has to be done one time. The budget is not set in stone. You may adjust the total budget amount and/or the materials cost according to students' math ability.

THE STANDARDS

SCIENCE

www.nextgenscience.org/search-standards-dci

The Next Generation Science Standards are arranged by disciplinary core ideas (DCI). When accessing these standards, search by standard and DCI. The standards are identified in the lessons by grade level and DCI (e.g., MS-ESS3-1–Grades 6-8, Earth and Human Activity, Standard 1).

TECHNOLOGY

www.iste.org/standards

The International Society for Technology in Education (ISTE) publishes the national technology standards. Each of the standards is categorized into four main categories.

1. Creativity and innovation
2. Communication and collaboration
3. Research and information fluency
4. Critical thinking, problem solving, and decision making

Within each of these categories there are more specific indicators that are identified by a letter. Standards within the lessons will be indicated by the category (e.g., ISTE.1).

ENGINEERING

www.nextgenscience.org/search-standards-dci

The Next Generation Science Standards identify the engineering standards as well. They are categorized by the grade band of 6-8 (e.g., MS-ETS1-1).

ARTS

www.nationalartsstandards.org
www.corestandards.org/ELA-Literacy

The National Core Arts Standards are divided into four categories:

1. Creating
2. Performing/Presenting/Producing
3. Responding
4. Connecting

Each of these categories contains anchor standards. Within the lesson, the standards will be identified by the category and the anchor standard (e.g., Creating, Anchor Standard #1).

In addition to performance standards, the literacy standards are embedded throughout the lessons. Each lesson identifies specific English language arts (ELA) standards (e.g., CCSS.ELA-LITERACY.W.8.2).

MATH

www.corestandards.org/math

The Common Core Math Standards are divided into two categories:

1. Content
2. Practice

The content standards are those items such as computation and geometry. The practice standards are a framework for ensuring that students are practicing math in a meaningful and appropriate manner.

The content standards will be identified first in the Math Standards column and the Math Practice Standards will be underneath (e.g., CCSS.MATH.CONTENT.8.G.A.2–real world graphing and CCSS.MATH.PRACTICE.MP4–model with mathematics).

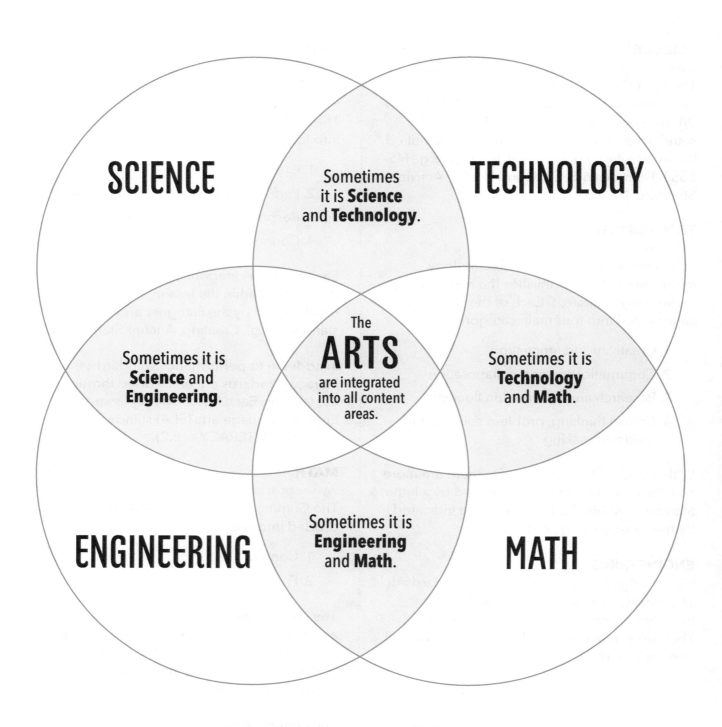

SCIENCE

Sometimes it is **Science** and **Technology**.

TECHNOLOGY

Sometimes it is **Science** and **Engineering**.

The **ARTS** are integrated into all content areas.

Sometimes it is **Technology** and **Math**.

ENGINEERING

Sometimes it is **Engineering** and **Math**.

MATH

Sometimes it is all five!

STEAM DESIGN PROCESS

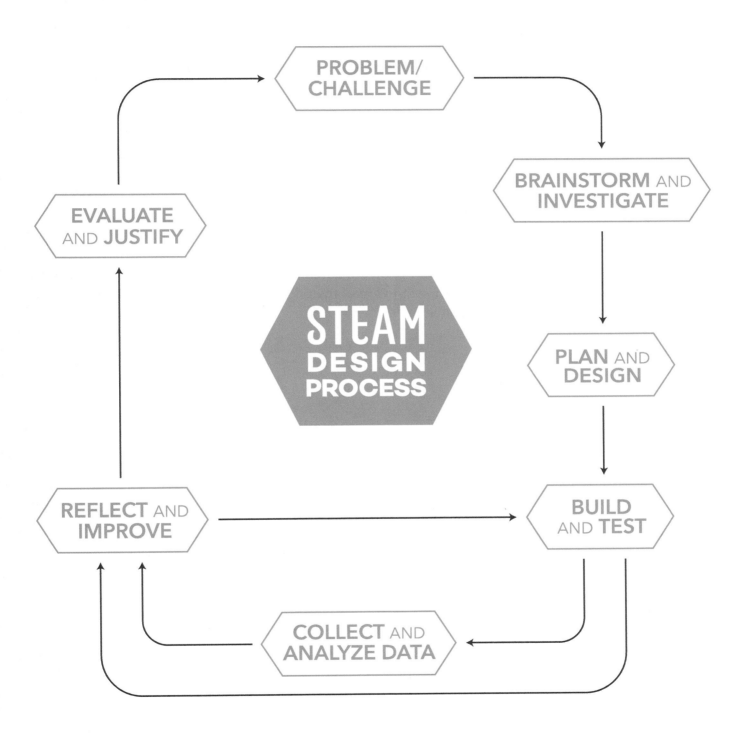

RECORDING INFORMATION IN A SCIENCE NOTEBOOK

Students will record their thinking, answer questions, make observations, and sketch ideas as they work through each design challenge. It is recommended that teachers have students designate a section of their regular science notebooks to these STEAM challenges or have students create a separate STEAM science notebook using a spiral notebook, a composition book, or lined pages stapled together.

Have students set up their notebooks based upon the natural breaks in the lesson. Remind students to write the name of the design challenge at the top of the page in their notebooks each time they prepare their notebooks for a new challenge.

Pages 1–3 Background Information

- Students record notes from any information provided by the teacher during whole-group instruction.

- Students record related vocabulary words and their definitions.

- Students record notes from their own independent research, including information gathered through literacy connections and existing background knowledge.

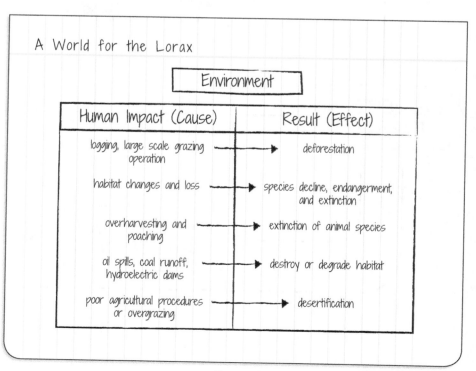

Page 1

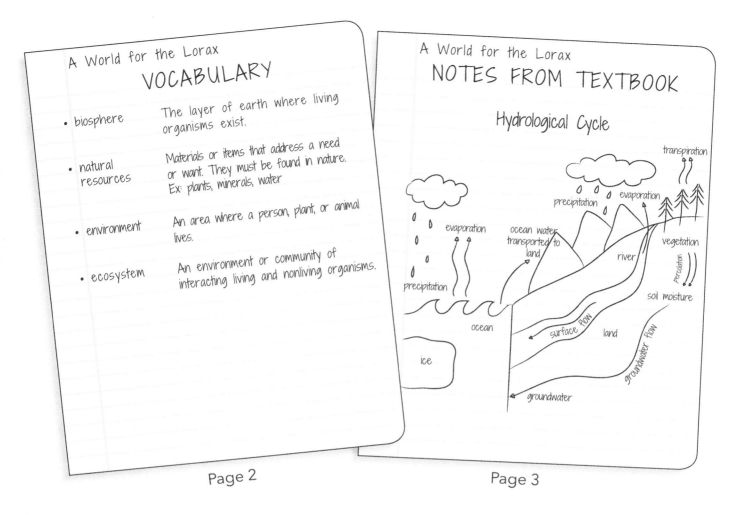

- biosphere — The layer of earth where living organisms exist.

- natural resources — Materials or items that address a need or want. They must be found in nature. Ex: plants, minerals, water

- environment — An area where a person, plant, or animal lives.

- ecosystem — An environment or community of interacting living and nonliving organisms.

A World for the Lorax
NOTES FROM TEXTBOOK

Hydrological Cycle

transpiration

precipitation

evaporation

evaporation

ocean water transported to land

precipitation

vegetation

river

percolation

soil moisture

ocean

surface flow

land

groundwater flow

ice

groundwater

Page 2

Page 3

Page 4 Dilemma and Mission

- Display the dilemma and mission for students to record.

- Or make copies of the dilemma and mission for students to glue into their notebooks to use as a reference.

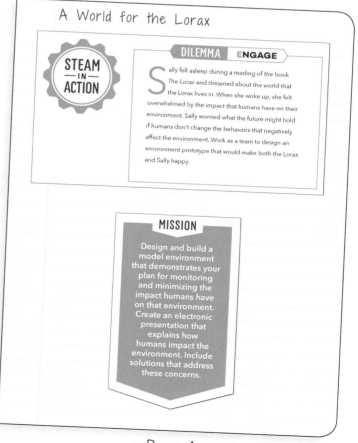

A World for the Lorax

STEAM IN ACTION

DILEMMA   ENGAGE

Sally fell asleep during a reading of the book *The Lorax* and dreamed about the world that the Lorax lives in. When she woke up, she felt overwhelmed by the impact that humans have on their environment. Sally worried what the future might hold if humans don't change the behaviors that negatively affect the environment. Work as a team to design an environment prototype that would make both the Lorax and Sally happy.

MISSION

Design and build a model environment that demonstrates your plan for monitoring and minimizing the impact humans have on that environment. Create an electronic presentation that explains how humans impact the environment. Include solutions that address these concerns.

Page 4

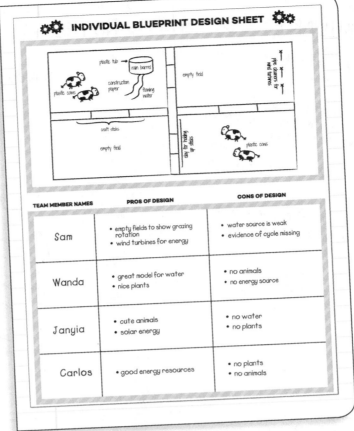

Page 5

Page 5 Blueprint Design

● Students draw their own suggested design. Then students write the pros and cons of both their and their teammates' designs.

● Or make copies of the Individual Blueprint Design Sheet for students to complete and glue into their notebooks.

REFLECTIONS EXPLAIN & ELABORATE

AFTER TEST TRIAL 1	What did your classmates like about your prototype and presentation? Summarize their feedback.
ANALYSIS	What will you do to improve your prototype?
AFTER TEST TRIAL 2	What observations did you make about your classmates' prototypes and presentations? Which prototype was most effective? Explain.
ANALYSIS	As a team, discuss how you can improve your prototype and presentation.
AFTER TEST TRIAL 3	According to the feedback from your classmates, did you successfully create a model environment that minimizes negative human impacts? Explain why you do or do not agree with their evaluations of your prototype.

Page 6

A World for the Lorax

TRIAL 1

My classmates really liked our use of solar energy to operate and create electricity for our world.

We felt we needed to present more information about water resources.

We are adding rain barrels and an irrigation system using the water from them. This will provide water for our plants without draining our aquifers.

TRIAL 2

Our favorite team created a forest of trees and a river. They showed people living in tree houses and using solar energy. It looked realistic.

We are going to redo our houses to make them more to scale with the model.

We are adding trees to provide shade and an animal habitat.

TRIAL 3

We felt our final model and presentation covered all the issues.

Our classmates were complimentary and we received good ratings.

Page 7

Pages 6–8 Engineering Task, Test Trial, Analyze, Redesign

- Students record analysis questions from the teacher and then record their answers. Or provide copies of the questions for students to glue into their notebooks.

- Record their reflections on the components of the prototypes that were successful and those that were not.

- Include additional pages as needed to allow students to record any notes, observations, and ideas as they construct and test their team prototype.

A World for the Lorax

SUMMARY

During this challenge, I learned about the many negative impacts that humans can have on the environment. My classmates and I proved that there are many ways to solve or minimize these impacts. My team and I used solar energy and rain barrels as part of an irrigation system. The rain barrels collect clean water, free from pollution. The water can be used for drinking as well as for watering plants. The model also showed our houses, which had solar panels. The solar energy can be used to provide electricity for the town. Overall, we did very well at meeting the needs of our town while minimizing the impact on the environment.

Page 8

A WORLD FOR THE LORAX

STEAMm

4-5 HOURS

TIME FOR COMPLETION

SETTING —THE— STAGE

DESIGN CHALLENGE PURPOSE

Design and construct a model environment, and create a plan for monitoring and minimizing the human impact on the environment.

Create an oral presentation that delivers information about your designed environment and its sustainability.

TEACHER DEVELOPMENT

Human activities have significantly altered the **biosphere**, sometimes damaging or destroying natural habitats and causing the extinction of other species. Changes to the environment can have both positive and negative impacts on living things. As human populations and consumption of **natural resources** increase, so does the negative impact on the earth unless humans take steps to prevent it.

For test trials, have student teams conduct a gallery walk of the projects, filling out the rubrics and leaving encouraging notes for other teams to discuss while redesigning their prototypes. For the trials, student teams give an oral presentation of their environment.

Note: Teachers should read *The Lorax* to students prior to the challenge but stop reading to begin the challenge after Once-ler's statement that nothing will get better unless someone cares a lot.

STUDENT DEVELOPMENT

Before beginning this challenge, students will need to access websites and other research materials to learn how humans impact their environment. Students will use this information to design their prototypes.

Students should examine how humans impact their **environment**, assessing the kinds of solutions that are feasible and designing and evaluating solutions that could reduce that impact.

Examples of human impacts can include water usage, such as the removal of water from streams and aquifers or the construction of dams and levees; land usage, such as urban development, agriculture, or the removal of wetlands; and pollution, such as pollution of the air, water, or land.

Note: Visit the website listed on the inside front cover for more information about how human activity affects the environment.

STANDARDS

SCIENCE	TECHNOLOGY	ENGINEERING	ARTS	MATH	ELA
MS-ESS3-3	ISTE.1	MS-ETS1-1	Creating #1		CCSS.ELA-LITERACY.W.6.3
MS-ESS2-4	ISTE.2	MS-ETS1-2	Performing #4		CCSS.ELA-LITERACY.W.6.8
	ISTE.3	MS-ETS1-3			
	ISTE.4	MS-ETS1-4			

SCIENCE & ENGINEERING PRACTICES

Developing and Using Models: Develop a model to describe unobservable mechanisms.

Constructing Explanations and Designing Solutions: Apply scientific principles to design, construct, and/or test a design of an object, tool, process, or system.

CROSSCUTTING CONCEPTS

Energy and Matter: Within a natural system, the transfer of energy drives the motion and/or cycling of matter.

Cause and Effect: Relationships can be classified as causal or correlational, and correlation does not necessarily imply causation.

Influence of Science, Engineering, and Technology on Society and the Natural World: The uses of technologies and any limitations on their use are driven by individual or societal needs, desires, and values; by the findings of scientific research; and by differences in such factors as climate, natural resources, and economic conditions.

Technology use varies over time and from region to region.

TARGET VOCABULARY

biosphere

environmental impact

hydrologic cycle

natural resources

transfer of energy

MATERIALS

- shoeboxes
- modeling clay
- construction paper
- pipe cleaners
- aluminum foil
- plastic wrap
- markers
- glue
- miscellaneous recyclable materials
- rubric (page 129)

LITERACY CONNECTIONS

The Lorax by Dr. Seuss

NOTES

STEAM IN ACTION

DILEMMA — ENGAGE

Sally fell asleep during a reading of the book *The Lorax* and dreamed about the world that the Lorax lives in. When she woke up, she felt overwhelmed by the impact that humans have on their environment. Sally worried what the future might hold if humans don't change the behaviors that negatively affect the environment. Work as a team to design an environment prototype that would make both the Lorax and Sally happy.

MISSION

Design and build a model environment that demonstrates your plan for monitoring and minimizing the impact humans have on that environment. Create an electronic presentation that explains how humans impact the environment. Include solutions that address these concerns.

BLUEPRINT — EXPLORE

Provide the Individual and Group Blueprint Design Sheets to engineering teams. Have individual students sketch a prototype to present to the other members of their team. Team members will discuss the pros and cons of each sketch and then select one prototype to construct.

ENGINEERING TASK **TEST TRIAL** **ANALYZE** **REDESIGN**

Each team will design and build a model environment that demonstrates the team's plan for monitoring and minimizing the impact humans have on their environment.

Teams will present their prototypes and record their observations.

Teams will present their models and electronic presentations to the class. They should reflect on the feedback from their peers using the rubric as a guide.

After analyzing the feedback, student teams review their original design. Teams then make adjustments and alter the original sketches using colored pencil to show the changes they have made. The goal is to improve their prototypes for the next trial.

HELPFUL TIPS

- After the Test Trial, have teams take a gallery walk to view other teams' designs for possible ideas to assist them in the Analyze and Redesign portions of the engineering design process.

- If teams are successful on the first try, encourage them to make their prototypes even more efficient. If it is a scenario in which this is not feasible, distribute team members to other teams to be a support for them in making their prototypes more efficient. Alternatively, at teacher discretion, move students on to the Justification portion of the lesson.

- If after the third test the final prototype is still unsuccessful, have students write how they would start over. These challenges are meant to have students build on what they originally designed. If the design proved to be unsuccessful, encourage a reflection or justification on what they would do if they were allowed to start again from scratch.

STEAM Design Challenges Gr. 6–8 © 2018 Creative Teaching Press

S T E A m

REFLECTIONS — EXPLAIN & ELABORATE

AFTER TEST TRIAL 1	What did your classmates like about your prototype and presentation? Summarize their feedback.
ANALYSIS	What will you do to improve your prototype?
AFTER TEST TRIAL 2	What observations did you make about your classmates' prototypes and presentations? Which prototype was most effective? Explain.
ANALYSIS	As a team, discuss how you can improve your prototype and presentation.
AFTER TEST TRIAL 3	According to the feedback from your classmates, did you successfully create a model environment that minimizes negative human impacts? Explain why you do or do not agree with their evaluations of your prototype.

JUSTIFICATION — EVALUATE

TECHNOLOGY	Create an electronic commercial designed to promote the protection of the environment.
ELA	Write an ending to the story *The Lorax* before hearing the end of the story.
ARTS	Design a flyer for an environmental protection agency.

HOW FAR MUST I GO?

STEAM

SETTING —THE— STAGE

DESIGN CHALLENGE PURPOSE

Create a heliocentric scale model of the orbital radius of the planets in our solar system.

TEACHER DEVELOPMENT

This lesson is designed to have students apply their understanding of scale and to deepen their conceptual understanding of each planet's orbital radius around the sun. In this challenge, **scale** refers to the determination the students make for how many kilometers the placement of their chosen object represents when creating their solar system prototypes. Our solar system is **heliocentric**. That simply means that a star (the sun) is at the center of our solar system. **Orbital radius** is the average distance between an orbiting planet and the sun. It's a great idea to have students guess the meaning of this term by breaking it apart. If they struggle, remind them that **orbit** means to follow a curved path and that **radius** is the distance from the center to the outside of a circle.

STUDENT DEVELOPMENT

Students will need to know that the solar system is heliocentric (centered on a star) and also know what celestial bodies exist in our solar system. Ensure that students are familiar with the vocabulary terms **heliocentric**, **scale**, **orbit**, **radius**, and **orbital radius**.

Allow students to use research materials to gather information about the solar system. Encourage them to record this information on note cards or in their science notebooks. Remind them to list the sources they use.

Note: Visit the website listed on the inside front cover for more information about the solar system.

STANDARDS

SCIENCE	TECHNOLOGY	ENGINEERING	ARTS	MATH	ELA
MS-ESS1-3	ISTE.1	MS-ETS1-1	Creating #1	CCSS.MATH. CONTENT.7.G.A.1	CCSS.ELA- LITERACY.W.7.7
MS-ESS1.A	ISTE.3	MS-ETS1-2	Creating #2	CCSS.MATH. CONTENT.7.G.B.4	CCSS.ELA- LITERACY.RST.6-8.7
	ISTE.6	MS-ETS1-3	Creating #3	CCSS.MATH. PRACTICE.MP2	
		MS-ETS1-4	Performing #5	CCSS.MATH. PRACTICE.MP4	

SCIENCE & ENGINEERING PRACTICES

Analyzing and Interpreting Data: Analyze and interpret data to determine similarities and differences in findings.

CROSSCUTTING CONCEPTS

Systems and System Models: Systems may interact with other systems; they may have sub-systems and be part of larger complex systems.

Models can be used to represent systems and their interactions—such as inputs, processes, and outputs—and energy and matter flows within systems.

Interdependence of Science, Engineering, and Technology: Engineering advances have led to important discoveries in virtually every field of science, and scientific discoveries have led to the development of entire industries and engineered systems.

TARGET VOCABULARY

celestial body

comet

heliocentric

orbital radius

planet

sun

MATERIALS

- items that can be used to show length for the scale model (like paper clips or cereal)
- toothpicks
- rubber bands
- sticky notes
- pretzel sticks
- licorice pieces
- chocolate chips
- bulletin board paper or large poster paper
- markers
- construction paper
- glue
- comment card (page 130)

LITERACY CONNECTIONS

Planets
by Ellen Hasbrouck

NOTES

STEAM —IN— ACTION

DILEMMA ENGAGE

Oliver Orbiter, chief executive officer of Outer Space Explorers, has a meeting with Sol Thermal, president of the Planetary Society, an organization that promotes space exploration. Oliver Orbiter wants the Planetary Society to fund his exploration of our solar system. Unfortunately, Oliver Orbiter's company recently suffered a cyber attack, and all of its data has been corrupted. It has taken him so long to prepare for the meeting, but he has no data to present. What Oliver does have is plenty of office supplies and snacks. Help him use the materials to create an eye-catching model of the solar system that he can use in his presentation.

MISSION

Research the distance from all of the planets in the solar system to the sun. Then create a scale model of the solar system representing the orbital radius of the planets. Include a scale that shows the distance in kilometers.

BLUEPRINT EXPLORE

Provide the Individual and Group Blueprint Design Sheets to engineering teams. Have individual students sketch a prototype to present to the other members of their team. Team members will discuss the pros and cons of each sketch and then select one prototype to construct.

Note: In this case, students will be deciding which materials work best for the model. They will also be defending their research and application of scale as it is applied to the distance from each planet to the sun. Students will not be making planet sizes to scale.

ENGINEERING TASK	TEST TRIAL	ANALYZE	REDESIGN
Teams will each create a scale model of the orbit of the planets around the sun. They will include a scale that shows the distance in kilometers represented by each paper clip or other item.	Teams will lay out their model solar system on the poster paper and then go on a gallery walk to observe and comment on other teams' models. All teams will complete a comment card after reviewing other teams' models. The comment card includes space to correct any inaccuracy they notice. Repeat this procedure for each trial. *Note*: If you do not wish to make copies of the reproducible comment cards, write the questions on the whiteboard and allow students to answer them on index cards or notebook paper.	Teams will review the comment cards they received from their classmates. They will check their notes and complete additional research if necessary. Teams will determine what they can do to improve their prototypes.	After analyzing the feedback, student teams review their original designs. Teams then make adjustments and alter the original sketches using colored pencil to show the changes they have made. The goal is to improve their prototypes for the next trial. Once teams are confident in their models, they may glue the pieces to the poster paper.

HELPFUL TIPS

- To encourage engagement, have students use food items in the construction of their models, or for a more budget-friendly route, use office supplies. For example, if they decide that one paper clip equals 500,000,000 km, the students would need 9 paperclips to lay out the measurement for the radius of Neptune's orbit, which is approximately 4.5 billion km from the sun.

REFLECTIONS — EXPLAIN & ELABORATE

AFTER TEST TRIAL 1	What did your classmates like about your prototype? Summarize their feedback regarding the accuracy of your scale model.
ANALYSIS	Based on the feedback from other teams, what changes will your team make to your model?
AFTER TEST TRIAL 2	What did you observe about the other teams' prototypes? Which one was your favorite? Explain why. Compare the accuracy and appearance of the different prototypes.
ANALYSIS	Discuss how you can improve your prototype.
AFTER TEST TRIAL 3	Based on the feedback your team received, did your team successfully create a scale model of each planet's orbital radius around the sun? Explain why you do or do not agree with the feedback you received.

JUSTIFICATION — EVALUATE

TECHNOLOGY	Create a slideshow presentation describing the planets in the solar system and their distances from the sun.
ELA	Research information about each planet and other celestial bodies in the solar system. Then use a computer publishing program to create a planet brochure featuring this information.
ARTS	Create a travel poster encouraging people to visit one planet other than Earth. Include the information you used to build your model.

WEATHER DETECTOR

SETTING —THE— STAGE

DESIGN CHALLENGE PURPOSE

Create a weather map of your state for one day. Include possible storms and high and low pressure fronts.

TEACHER DEVELOPMENT

Airplane flights and cruise ships frequently get redirected to avoid large thunderstorms. This data is available on several websites to view and print. You will need to obtain flight data for several flights arriving and departing from the major airports in your state. Print the flight data, but do not include the flight map that shows the weather. The data needed will show planes rerouting around large air masses. You will give the printed data to the teams to use as they create their weather maps. If your state borders an ocean, you can also collect cruise ship data to print and give to student teams.

STUDENT DEVELOPMENT

Students will need to have an understanding of weather maps and their symbols. Background knowledge of cold, warm, and stationary fronts—and how they interfere with air and cruise ship travel—is also important.

A **cold front** is the advancing edge of a cold air mass, and a **warm front** is the advancing edge of a warm air mass. A **stationary front** is the boundary between two air masses of which neither is replacing the other.

Note: Visit the website listed on the inside front cover for more information about flight data.

STANDARDS

SCIENCE	TECHNOLOGY	ENGINEERING	ARTS	MATH	ELA
MS-ESS2-5		MS-ETS1-1	Creating #1	CCSS.MATH. CONTENT.6.SP.B.4	CCSS.ELA-LITERACY.RST.6-8.1
		MS-ETS1-2		CCSS.MATH. CONTENT.6.SP.B.5	
		MS-ETS1-3			
		MS-ETS1-4			

SCIENCE & ENGINEERING PRACTICES

Planning and Carrying Out Investigations: Collect data to produce data to serve as the basis for evidence to answer scientific questions or test design solutions under a range of conditions.

CROSSCUTTING CONCEPTS

Cause and Effect: Cause-and-effect relationships may be used to predict phenomena in natural or designed systems.

TARGET VOCABULARY

air masses
analyze
flight data
fronts
high pressure
humidity
low pressure
precipitation
pressure
temperature
wind

MATERIALS

- colored pencils
- flight data
- cruise ship data (if applicable)
- blank map of your state

LITERACY CONNECTIONS

Mapping the Land and Weather
by Melanie Waldron

NOTES

DILEMMA — ENGAGE

A national TV weather news station has lost contact with its weather satellites, and it needs information from the satellites to broadcast its latest weather reports. The only data that can be accessed is that of plane flights, cruise ships, and temperature readings from major cities. The news station is reaching out to each state, asking teams to create state weather maps that it can use to make a weather map of the United States. Can you help the television station by creating a weather map of your state?

MISSION

Create a weather map of your state showing any storms and high or low pressure fronts. Then develop a weather forecast for your town.

BLUEPRINT — EXPLORE

Provide the Individual and Group Blueprint Design Sheets to engineering teams. Have individual students sketch a prototype to present to the other members of their team. Team members will discuss the pros and cons of each sketch and then select one prototype to construct.

ENGINEERING TASK	TEST TRIAL	ANALYZE	REDESIGN
Teams will each construct a weather map of their state and create a report for their town.	Teams will display their weather maps for review by the other teams and the teacher.	After reviewing the other teams' maps, team members can discuss the similarities and differences between the different maps and then reevaluate their data and map design.	After analyzing the feedback, student teams review their original designs. Teams then make adjustments and alter the original sketches using colored pencil to show the changes they have made. The goal is to improve their prototypes for the next trial.

HELPFUL TIPS

- After the Test Trial, have teams take a gallery walk to view other teams' designs for possible ideas to assist them in the Analyze and Redesign portions of the engineering design process.

- If teams are successful on the first try, encourage them to make their prototypes even more efficient. If it is a scenario in which this is not feasible, distribute team members to other teams to be a support for them in making their prototypes more efficient. Alternatively, at teacher discretion, move students on to the Justification portion of the lesson.

- If after the third test the final prototype is still unsuccessful, have students write how they would start over. These challenges are meant to have students build on what they originally designed. If the design proved to be unsuccessful, encourage a reflection or justification on what they would do if they were allowed to start again from scratch.

STEAM Design Challenges Gr. 6–8 © 2018 Creative Teaching Press

S t E A m

REFLECTIONS — EXPLAIN & ELABORATE

AFTER TEST TRIAL 1	What were some of the similarities and differences between your team's weather map and the maps belonging to the other teams?
ANALYSIS	What data might your team need to review again to determine whether or not you must make any changes to your map? Did your map include air masses and fronts?
AFTER TEST TRIAL 2	After viewing the other teams' updated maps, what improvements did you notice? What changes did you make to your own map?
ANALYSIS	What were some of the differences between your team map and the other maps? What data might your team need to review again to determine whether or not you must make any changes to your map?
AFTER TEST TRIAL 3	What was the most difficult part of this challenge? Did your team successfully create a weather map?

JUSTIFICATION — EVALUATE

ELA	Using data from a real weather map, create a flight plan from one US city to another. Include a written justification for the plan you created.
ARTS	Perform a meteorology news segment. Include appropriate vocabulary terms and accurate weather props.
MATH	Select a city that is far from your home city, and research daily average temperatures and weather. Display the final data.

SHIFTING MINES

STEAM

SETTING —THE— STAGE

DESIGN CHALLENGE PURPOSE

Analyze and interpret data from fossils, rocks, continental shapes, and seafloor structures to create a labeled 3-D model that provides evidence that South America was once connected to Africa by land.

TEACHER DEVELOPMENT

This lesson builds on students' understanding of the **law of superposition** (the basic understanding that when looking at layers of rock, the newest layers are at the top and the oldest layers are at the bottom) and the **cross-cutting principle** (when a younger rock cuts across several layers of older rock). They should also know about Alfred Wegener's theory of **continental drift**. You can familiarize students with continental drift by reading aloud Wegener's "Building a Case for Continental Drift."

Note: Visit the website listed on the inside front cover to show students a map that illustrates the possible connection between diamonds on separate continents, as discussed in the Dilemma section. The map is extremely beneficial to understanding the connections of diamonds and separate lands discussed in the Dilemma section.

STUDENT DEVELOPMENT

Geologists use both relative and absolute age to determine the age of rocks and fossils. **Relative age** is the age of a rock compared to other rocks, and **absolute age** is how long ago the rock was formed. Sedimentary rocks are formed in layers. The youngest rocks are on the top layer, and the older rocks were formed in sequence in the layers below. When a rock intrudes into another rock, the **intrusive** rock is younger in age than the rock that it intruded into. When a rock is formed by **extrusive** means, it is younger than the rock it forms on. Scientists also use index fossils to help date rocks. **Index fossils** are fossils that are found in a large variety of places and the animal or plant only lived a short period of time geologically.

STANDARDS

SCIENCE	TECHNOLOGY	ENGINEERING	ARTS	MATH	ELA
MS-ESS2-3	ISTE.3a	MS-ETS1-1	Creating #1	CCSS.MATH. CONTENT.6.SP.B.4	CCSS.ELA-LITERACY.RST.6-8.1
	ISTE.3b	MS-ETS1-2	Performing #5	CCSS.MATH. CONTENT.6.SP.B.5	CCSS.ELA-LITERACY.RST.6-8.7
	ISTE.3c	MS-ETS1-3			
	ISTE.3d	MS-ETS1-4			

SCIENCE & ENGINEERING PRACTICES

Analyzing and Interpreting Data: Analyze and interpret data to provide evidence for phenomena.

CROSSCUTTING CONCEPTS

Patterns: Patterns in rates of change and other numerical relationships can provide information about natural systems.

TARGET VOCABULARY

continental shelf

continental slope

cross-cutting principle

extrusive

fracture zone

index fossil

intrusive

law of superposition

plate tectonics

ridge

trench

MATERIALS

- paper
- clay
- toothpicks
- recycled items that students bring from home
- rubric (page 131)

Note: This is an ideal project for a makerspace learning center.

LITERACY CONNECTIONS

Fault Lines & Tectonic Plates: Discover What Happens When the Earth's Crust Moves by Kathleen M. Reilly

What Is the Theory of Plate Tectonics? by Craig Saunders

NOTES

STEAM
—IN—
ACTION

DILEMMA — ENGAGE

Amil Allotrope, president of West African Diamond Mines, has a problem. His mines are running low on diamonds, and he needs to find another deposit so the company can mine more diamonds. He has a hunch that they should dig farther west, along the coast of western Africa, or maybe even buy land in eastern Brazil. In order to do this, he needs to convince the West African Diamond Mines board of governors that his hunch is correct. He needs a team of researchers to find the proof that the company should buy land farther west or in eastern Brazil. Can your team help Mr. Allotrope by creating a model that supports the theory that South America was once connected to Africa by land?

MISSION

Create a 3-D model to help prove that South America was once connected to Africa by land. Project requirements:

1. Data showing evidence of similar index fossils on both continents.

2. Data showing evidence of similar rocks on both continents.

3. Data showing the location of diamonds on both continents.

4. Continental shapes, including continental shelves and slopes.

5. Labeled seafloor structures.

BLUEPRINT — EXPLORE

Provide the Individual and Group Blueprint Design Sheets to engineering teams. Have individual students sketch a prototype to present to the other members of their team. Team members will discuss the pros and cons of each sketch and then select one prototype to construct.

ENGINEERING TASK	**TEST TRIAL**	**ANALYZE**	**REDESIGN**
Teams will each construct a 3-D model that has labeled evidence to support the theory that South America and Africa were once connected by land. The model must include the five mission requirements and cite textual support.	Teams will display their prototypes for a class gallery walk, where students will observe and give feedback about other teams' designs. *Note:* After two class gallery walks, the final trial model can be evaluated by the teacher.	Teams should analyze the feedback from the other teams. Teams should also be given the rubric to use when reviewing their own design.	After analyzing the feedback from the two gallery walks, student teams will review their most recent design and make changes in preparation for the final evaluation by the teacher. Teams should make adjustments and alter the original sketches using colored pencil to show the changes they have made.

 HELPFUL TIPS

- After the Test Trial, have teams take a gallery walk to view other teams' designs for possible ideas to assist them in the Analyze and Redesign portions of the engineering design process.

- If teams are successful on the first try, encourage them to make their prototypes even more efficient. If it is a scenario in which this is not feasible, distribute team members to other teams to be a support for them in making their prototypes more efficient. Alternatively, at teacher discretion, move students on to the Justification portion of the lesson.

- If after the third test the final prototype is still unsuccessful, have students write how they would start over. These challenges are meant to have students build on what they originally designed. If the design proved to be unsuccessful, encourage a reflection or justification on what they would do if they were allowed to start again from scratch.

REFLECTIONS — EXPLAIN & ELABORATE

AFTER TEST TRIAL 1	Did your team's prototype meet all of the requirements listed on the rubric?
ANALYSIS	Look at the feedback from the other teams and determine what changes you will make to your team's design.
	Based on the rubric evaluation, what are your prototype's strengths and weaknesses?
AFTER TEST TRIAL 2	How did the changes your team made help improve your prototype?
ANALYSIS	Look at the feedback from the second review and determine if your team will make any final improvements.
	What changes did your team decide to make to your original prototype?
AFTER TEST TRIAL 3	Did your team's prototype meet all of the requirements on the rubric?
	What changes did you make to improve your team's rubric score?

JUSTIFICATION — EVALUATE

TECHNOLOGY	Create an electronic presentation that explains your prototype.
ELA	Write a brief report that summarizes the data you collected during your research.
ART	Create another display showing other continents and their similarities.
MATH	Display the data you collected during your research in a histogram or other appropriate graphical display (e.g., flow chart, graph, table).

WHO DONE IT?

2-8 HOURS

*TIME FOR COMPLETION

STEAM

SETTING —THE— STAGE

DESIGN CHALLENGE PURPOSE

Design two escape room games for the Who Done It Escape House.

TEACHER DEVELOPMENT

Escape rooms are very popular now as a form of entertainment. Many cities offer them, and some establishments even offer rooms for children to solve puzzles. Some popular themes include the White House, zombies, classrooms, spy headquarters, and crime scenes. Participants pay to go into the rooms and use clues to try to solve many puzzles to open locks and latches before time is up. The games are thrilling because it is a race against the clock to solve the puzzles and get out before the time runs out. Most games are designed to be difficult, and only a small percentage of participants typically complete them.

Encourage students to create patterns for game participants to recognize in the clues. For example, if the theme is a zoo, the clues could all be related to animals. Encourage students to think of recent science and math concepts to apply to their clues. For example, if they just studied geometry, students could design a puzzle using 3-D shapes. When the shapes are constructed, they could have a clue spelled out for the next puzzle.

Half of the fun of this challenge is creating the room. The other half is trying to solve the puzzles in other teams' rooms. You can have students try their classmates' rooms, or partner with another classroom and have your students try the other students' game rooms, and then rate them. Students will get important feedback about their game, similar to what a real business would receive.

*See the Helpful Tip on page 48 for information about the extended lesson length.

STUDENT DEVELOPMENT

Students will need ideas about how to build their clues. To stimulate their thinking, include materials related to forces such as magnets, electrical wiring, light switches, letter tiles, dominoes, cards, blank crossword puzzles, familiar poems, and word games. Students should also recall their knowledge of Newton's third law of motion.

Have students select the theme of their escape room. For example, if the theme is a zoo, participants might have to solve a crossword puzzle about animals and each clue answer would have one letter circled. After they complete the puzzle, they take the circled letters and figure out the word they spell out. The word would give a clue for the next question.

Note: Visit the website listed on the inside front cover to show students an example of what an escape room is all about and for an example of a completed puzzle.

STANDARDS

SCIENCE	TECHNOLOGY	ENGINEERING	ARTS	MATH	ELA
MS-PS2-1	ISTE.3	MS-ETS1-1	Creating #1	CCSS.MATH.CONTENT.6.SP.A.1	CCSS.ELA-LITERACY.WHST.6-8.2
MS-PS2-3		MS-ETS1-2	Creating #2		
MS-PS2-5		MS-ETS1-3	Creating #3		
		MS-ETS1-4			

SCIENCE & ENGINEERING PRACTICES

Asking Questions and Defining Problems: Ask questions that can be investigated with available resources within the scope of the classroom, an outdoor environment, or a public facility, such as a museum. When appropriate, frame a hypothesis based on observations and scientific principles.

Planning and Carrying Out Investigations: Plan an investigation individually and collaboratively. In the design, identify independent and dependent variables and controls, what tools are needed to do the gathering, how measurements will be recorded, and how many data are needed to support a claim. Conduct an investigation and/or evaluate and/or revise the experimental design to produce data to serve as the basis for evidence that can meet the goals of the investigation.

Constructing Explanations and Designing Solutions: Apply scientific ideas or principles to design, construct, and/or test a design of an object, tool, process, or system.

CROSSCUTTING CONCEPTS

Cause and Effect: Cause-and-effect relationships may be used to predict phenomena in natural or designed systems.

Systems and System Models: Models can be used to represent systems and their interactions—such as inputs, processes, and outputs—and energy and matter flows within systems.

Stability and Change: Explanations of stability and change in natural or designed systems can be constructed by examining the changes over time and processes at different scales, including the atomic scale.

TARGET VOCABULARY

electrical force

force

gravitational force

magnetic attraction

magnetic force

repulsion

MATERIALS

- graph paper
- printer paper
- pencil
- ruler
- puzzle design sheet (page 132)
- rating sheet (page 133)
- Budget Planning Chart (page 145)

Other suggestions:

- electrical wires
- magnets
- steel nails
- switches
- letter tiles
- puzzles
- domino tiles

Note: If students need more room to write, enlarge copies of the puzzle design sheet and print on 11" x 17" paper (zoom ratio 1.294 or 129%).

LITERACY CONNECTIONS

How to Create a Low Cost Escape Room: For Camps, Youth Groups and Community Centers
by Curt "Moose" Jackson

Paper Escapes: A Fun and Exciting ESCAPE ROOM Experience at Home
by Jesse Cruz

NOTES

DILEMMA ENGAGE

Mr. Andrew Locksmith owns an escape room arcade that has lots of customers. Mr. Locksmith's Who Done It Escape House is a very popular place. It is so popular, in fact, that most of the townspeople have already visited the current rooms and solved, or tried to solve, all of the puzzles. To keep his arcade popular, he must stay current and needs more rooms designed. He needs your help. Can you pick a theme and come up with clues for that theme in order to create two new escape rooms? Mr. Locksmith has a limited budget for creating the new rooms, so you will need to choose your materials wisely in order to stay within the budget.

MISSION

Requirements:

1. Create two rooms with the same theme.

2. Each room must have at least five problems to solve and at least two unique locks or latches.

3. One problem must demonstrate Newton's third law of motion.

4. One problem must demonstrate electric, magnetic, or gravitational force.

5. One problem must be mathematical and solve a statistical problem.

6. Write game rules that allow players to ask for help twice.

BLUEPRINT EXPLORE

Have team members use the Blueprint Design Sheets to plan the placement of challenges within their room. Also give students a copy of the puzzle design sheet when designing puzzle ideas they want to include in their team's escape room. Students will present their ideas to their teams. Have students list pros and cons for each team member's suggestions. Before teams purchase materials, they must write why they selected their puzzles and why they feel that the other designs would not work as well. They may include all of their challenge ideas and more if the budget allows.

Note: You may set the budget amount and material costs according to your students' level of math abilities.

ENGINEERING TASK   **TEST TRIAL**   **ANALYZE**   **REDESIGN**

ENGINEERING TASK	TEST TRIAL	ANALYZE	REDESIGN
Teams will use the materials provided to create their puzzles and clues. They should refer to their puzzle design sheet when building their puzzles.	Test Trial 1: Teams will test their own rooms and complete a rating sheet to ensure the puzzles make sense and can be solved in the time allotted by the teacher. They will make changes based on the rating sheet results. Test Trial 2: Teams will solve another team's rooms and fill out a rating sheet. Then they will make changes to their own design based on the feedback their rooms received from the other team. Test Trial 3: Teams will solve a new team's rooms and fill out a rating sheet. Once all rooms have been rated by two other teams, the team with the best score on the third rating sheet wins the challenge.	Teams must complete a rating sheet after each trial. They should review their results after the first two trials to determine what they need to change. The third rating sheet will determine the escape game winner.	Teams use the feedback from the rating sheet to decide how to redesign their rooms, to determine if the puzzles worked, and to decide if the puzzles were challenging enough. Teams can use the puzzle design sheet to rework their designs. Their new design must also have an updated budget sheet with the correct calculations prior to redesigning their puzzles.

HELPFUL TIPS

- The time needed for this challenge varies depending on the needs of your students. If you choose to have teams create one puzzle each day, they should have their puzzle room(s) outlined by the end of the week.

REFLECTIONS — EXPLAIN & ELABORATE

AFTER TEST TRIAL 1	Solve the clues in your team's rooms to ensure all of the clues make sense. Fill out a rating sheet. Discuss the results and make changes if necessary.
ANALYSIS	Does your sequence of activities make sense? Did you leave out an important clue? What adjustments can you make to follow your theme and challenge participants?
AFTER TEST TRIAL 2	Have another team fill out a rating sheet as its members test your rooms by solving the clues. Review the results of the rating sheet and make changes if necessary.
ANALYSIS	Did the other team get stuck and need help? If so, what changes can you make to your design to make it more understandable yet still fun and challenging? Will your game room be selected for *Who Done It Escape House*?
AFTER TEST TRIAL 3	Have another team fill out a rating sheet as its members test your rooms by solving the clues. Review the results of the rating sheet. Which team had the highest score on its third rating sheet?

JUSTIFICATION — EVALUATE

TECHNOLOGY	A. Work with your teacher to create a digital blueprint of the game. B. Create posters, logos, and rule sheets for your game. Use an online crossword puzzle creator to create two additional puzzles for your rooms.
ELA	Explain in great detail how you created your escape room.
ARTS	Design an advertisement for your escape room.

ENGINEER THAT!

3-4 HOURS

TIME FOR COMPLETION

S T E A m

SETTING
—THE—
STAGE

DESIGN CHALLENGE PURPOSE

Build a Rube Goldberg machine that moves a load across a distance without causing harm.

TEACHER DEVELOPMENT

This lesson will require students first to do some research on environmentally friendly mining. Mining can become more eco-friendly by developing and integrating practices that reduce the environmental impact of mining operations. Some examples include reducing water and energy consumption, minimizing land disturbance and waste production, preventing soil, water, and air pollution at mine sites, and conducting successful mining activities.

Students will then need to understand what a Rube Goldberg machine is. Rube Goldberg machines are devices featuring uniquely over-engineered chain reactions.

Note: Visit the website listed on the inside front cover for information about environmentally friendly mining.

STUDENT DEVELOPMENT

While mining has historically affected its surrounding environment, advances in technology and changes in mining techniques mean that many of its negative impacts are now avoidable. Increasingly, mining companies are making efforts to reduce their environmental impact and minimize the footprint of their activities. This includes working to restore ecosystems post-mining.

Students must use what they have learned about sustainable mining as well as background information on Rube Goldberg machines to complete this challenge.

Students will also need to recall the basics of magnetic attraction and repulsion as well as different types of energy transfers, and Newton's laws of motion.

Note: Visit the website listed on the inside front cover for more information, including a video, about Rube Goldberg machines.

STANDARDS

SCIENCE	TECHNOLOGY	ENGINEERING	ARTS	MATH	ELA
MS-PS2-1	ISTE.3	MS-ETS1-1	Creating #1		CCSS.ELA-LITERACY.SL.6.1
MS-PS2-2		MS-ETS1-2	Creating #2		CCSS.ELA-LITERACY.W.6.3
MS-PS2-3		MS-ETS1-3	Creating #3		
MS-PS2-4		MS-ETS1-4			
MS-PS2-5					

SCIENCE & ENGINEERING PRACTICES

Asking Questions and Defining Problems: Define a design problem that can be solved through the development of an object, tool, process, or system and includes multiple criteria and constraints, including scientific knowledge that may limit possible solutions.

Developing and Using Models: Develop and/or use a model to generate data to test ideas about phenomena in natural or designed systems, including those representing inputs and output and those at unobservable scales.

Analyzing and Interpreting Data: Analyze and interpret data to determine similarities and differences in findings.

Engaging in Argument from Evidence: Evaluate competing design solutions based on jointly developed and agreed-upon design criteria.

CROSSCUTTING CONCEPTS

Influence of Engineering, Technology, and Science on Society and the Natural World: All human activity draws on natural resources and has both short- and long-term consequences, positive as well as negative, for the health of people and the natural environment.

The uses of technologies and any limitations on their use are driven by individual or societal needs, desires, and values; by the findings of scientific research; and by differences in such factors as climate, natural resources, and economic conditions.

TARGET VOCABULARY

attract

balanced force

cause

effect

force

motion

patterns

repel

unbalanced force

MATERIALS

- cups
- paper towel rolls
- plates
- craft sticks
- string
- tape

Optional:
toy and game parts (e.g., balls, blocks, cards, dominoes, marbles, and toy race track parts), various materials students bring from home

LITERACY CONNECTIONS

Ruby Goldberg's Bright Idea
by Anna Humphrey

The Best of Rube Goldberg
by Reuben Lucius Goldberg

NOTES

STEAM
— IN —
ACTION

DILEMMA ENGAGE

The Environmental Protection Agency (EPA) is concerned with the amount of pollution and environmental damage caused by mining across the globe. Mine waste, including solid waste, mine water, and fine particles have a high potential for environmental contamination. The EPA is looking for a machine that can move mine waste to an appropriate storage facility without harming the environment. The machine must maneuver around natural barriers without causing additional damage to the landscape. Can your team build a machine for the EPA?

MISSION

Design and construct a device that operates like a Rube Goldberg machine.

Device Requirements:

1. Must have two cause-and-effect relationships that demonstrate Newton's first law of motion.

2. Must have an electromagnet that either attracts or repels another magnet.

3. Must finish by demonstrating Newton's third law of motion.

4. Must be at least 5 feet in length.

BLUEPRINT EXPLORE

Provide the Individual and Group Blueprint Design Sheets to engineering teams. Have individual students sketch a prototype to present to the other members of their team. Teams will discuss the pros and cons of each sketch and then select one prototype to construct.

ENGINEERING TASK   **TEST TRIAL**   **ANALYZE**   **REDESIGN**

ENGINEERING TASK	TEST TRIAL	ANALYZE	REDESIGN
Teams will each design and construct a Rube Goldberg machine that includes two cause-and-effect relationships that demonstrate Newton's first law of motion; has an electro-magnet that attracts or repels another magnet; finishes by demonstrating Newton's third law of motion; and is at least 5 feet long.	Teams will test their Rube Goldberg machines, noting where they need to make adjustments. There may need to be more than three trials before teams are completely successful. Teams need to be sure to meet all of the requirements for their machines to pass the test. *Note:* Instruct the students that if a team's machine stops working during a test trial, they should keep going by picking up from the next part of the machine. This will keep them focused on taking notes and collecting data for their next trials.	Teams will need to take note of where their machines stop working and then determine where to make adjustments.	Teams will use the notes that they collected during their test trials to decide how to adjust their designs.

HELPFUL TIPS

- After the Test Trial, have teams take a gallery walk to view other teams' designs for possible ideas to assist them in the Analyze and Redesign portions of the engineering design process.

- If teams are successful on the first try, encourage them to make their prototypes even more efficient. If it is a scenario in which this is not feasible, distribute team members to other teams to be a support for them in making their prototypes more efficient. Alternatively, at teacher discretion, move students on to the Justification portion of the lesson.

- If after the third test the final prototype is still unsuccessful, have students write how they would start over. These challenges are meant to have students build on what they originally designed. If the design proved to be unsuccessful, encourage a reflection or justification on what they would do if they were allowed to start again from scratch.

REFLECTIONS — EXPLAIN & ELABORATE

AFTER TEST TRIAL 1	Did your machine have at least two cause-and-effect relationships that demonstrate Newton's first law of motion? Did it have an electromagnet that attracted or repelled a magnet? Did the machine finish by demonstrating Newton's third law of motion? Was your machine at least 5 feet long? Did your machine move the load completely through each task?
ANALYSIS	Did you have any malfunctions in your machine? Do you need to adjust a component of your machine in order to complete the challenge?
AFTER TEST TRIAL 2	Did all of your adjustments work successfully? Did you meet all of the requirements of the challenge?
ANALYSIS	Do you need to make any adjustments in order to meet all of the requirements?
AFTER TEST TRIAL 3	Did your machine work successfully? Did your machine meet all the requirements of the challenge? What could you add to your machine in the future?

JUSTIFICATION — EVALUATE

TECHNOLOGY	Record your Rube Goldberg machine in action! Describe what is happening at each stage. Post the final video on the class website.
ELA	Write a story about the trip that a small character takes as he or she travels through your Rube Goldberg machine.
ARTS	Create a Rube Goldberg cartoon strip featuring your prototype.

RIVETING ROLLER COASTERS

3-4 HOURS

TIME FOR COMPLETION

SETTING
—THE—
STAGE

DESIGN CHALLENGE PURPOSE

Design, construct, and test a roller coaster with at least two loops.

TEACHER DEVELOPMENT

In this challenge, students will be testing the relationship between **gravitational potential energy** (the amount of energy an object has due to its position and the gravitational pull acting on it) and the average speed of an object as the potential energy transforms to **kinetic energy** (the energy of motion). **Average speed** measures the rate of change in distance as it relates to time. Teams will use the formula Speed = Distance/Time to determine the average speed of the marble on their roller coasters.

You will find the math differentiated by grade levels in the Justification section of the lesson. In order to complete the math justification, students will be asked to collect and record data during their test trials. Although it will be used differently by different grade levels, the collecting and recording portion that occurs during the test trials is the same.

STUDENT DEVELOPMENT

Ensure that students are familiar with the scientific vocabulary needed for this challenge. Build background for your students by discussing the mechanics and origin of roller coasters. Unlike many modern steel-constructed roller coasters, the original roller coasters were made of wood. Many historians believe that the very first roller coasters were based on Russian slides, which were hills of ice. From there, historians differ as to whether the first real roller coaster was constructed in Russia or France. What is known is that the roller coaster has become faster, higher, and more thrilling over time.

Lesson Idea: Place students into groups of two. Working with a partner, have students prop up one end of a textbook approximately 5 centimeters off the desk to create a ramp. Have groups hold a ping-pong ball at the highest end of the ramp. Using a stopwatch, the groups will time and record how long it takes for the ball to move from the top of the ramp to the bottom. Have groups measure the length of the textbook ramp. Then have them use the formula S=D/T to find the speed of the ball. As an extension, groups can raise and lower the height of the ramp to see how the speed of the ball changes depending on the height of the ramp.

STANDARDS

SCIENCE	TECHNOLOGY	ENGINEERING	ARTS	MATH	ELA
MS-ESS3-2		MS-ETS1-1	Creating #1	CCSS.MATH.CONTENT.6.SP.B.5	CCSS.ELA-LITERACY.WHST.6-8.2
		MS-ETS1-2	Creating #2	CCSS.MATH.CONTENT.7.SP.A.2	
		MS-ETS1-3	Creating #3	CCSS.MATH.CONTENT.8.SP.A.1	
		MS-ETS1-4			

SCIENCE & ENGINEERING PRACTICES

Constructing Explanations and Designing Solutions: Apply scientific ideas or principles to design, construct, and/or test a design of an object, tool, process, or system.

CROSSCUTTING CONCEPTS

Cause and Effect: Cause-and-effect relationships may be used to predict phenomena in natural or designed systems.

TARGET VOCABULARY

gravitational potential energy

height

kinetic energy

roller coaster

speed

MATERIALS

- 1½" x 6' foam pipe insulation (cut in half lengthwise to make two 6' lengths, then taped together to create one 12'-long track)
- marble (to test the roller coaster)
- duct tape
- stopwatch
- scissors

LITERACY CONNECTIONS

50 Groundbreaking Roller Coasters: The Most Important Scream Machines Ever Built by Nick Weisenberger

Roller Coaster: Wooden and Steel Coasters, Twisters and Corkscrews by David Bennett

NOTES

S t E A M

STEAM —IN— ACTION

DILEMMA   ENGAGE

Mr. Rick Roboto is determined to make his Rockin' Rollin' Robotics amusement park absolutely riveting. He wants to make an automated roller coaster. Customers will be able to choose the speed of their ride by choosing the starting height for their first drop. He knows this would draw in hundreds of new customers, but he's not sure how to build this remarkable roller coaster. He needs your help to create a roller coaster prototype.

MISSION

Design, build, and test a roller coaster that makes two full loops. Once your prototype is successful, calculate the speed of your roller coaster when the starting point is set at three different heights.

BLUEPRINT   EXPLORE

Provide the Individual and Group Blueprint Design Sheets to engineering teams. Have individual students sketch a prototype to present to the other members of their team. Teams will discuss the pros and cons of each sketch and then select one prototype to construct.

ENGINEERING TASK

 TEST TRIAL

 ANALYZE

REDESIGN

Each team will design and construct a roller coaster that makes at least two loops. Once the marble successfully travels through the roller coaster, calculate the speed of the roller coaster when the starting point is set at three different heights. Each of the three heights must be at least 5 centimeters apart. Test each starting height five times.

Teams will use duct tape to secure the starting point of the roller coaster to a solid surface. Teams must measure and record the height of the starting point from the ground.

The marble must travel the entire length of the roller coaster unassisted. Teams will record the time it takes for the marble to reach the end of the track. Teams should collect data for the speed of the marble released at each of the three heights five times before changing the height. Each of the three heights must be at least 5 centimeters apart.

The speed of the marble will be calculated using the equation:

$$\text{speed} = \frac{\text{distance}}{\text{time}}$$

Facilitate analytical discussions comparing height, speed, and the designs of the different roller coaster prototypes. Allow teams to reflect on their design compared to others and how they might change their team's design.

Students will return to their design and use colored pencil to make changes based on their observations.

 HELPFUL TIPS

- Set a total budget for the challenge, and set a price for each item that students want to use to build their prototypes. Limiting students with a budget simulates a real-world problem. The budget total is given to students with the understanding that they cannot exceed that amount for all three of their tests.

STEAM Design Challenges Gr. 6–8 © 2018 Creative Teaching Press

REFLECTIONS — EXPLAIN & ELABORATE

AFTER TEST TRIAL 1	What characteristics did the most successful prototypes have in common? Was your team successful? Explain your observations.
ANALYSIS	What changes will your team make to your prototype? Explain the reason for these changes.
AFTER TEST TRIAL 2	Was your roller coaster successful? Which team had the fastest roller coaster? What was the team's greatest speed?
ANALYSIS	What changes will you make to your prototype? Explain the reason for these changes.
AFTER TEST TRIAL 3	What was the fastest recorded speed? What was the slowest recorded speed?

JUSTIFICATION — EVALUATE

ARTS	Create a theme park map that shows the locations of all the rides at the Rockin' Rollin' Robotics amusement park, including your prototype roller coaster.
MATH GRADE 6	Summarize the data set collected from your test trials by finding the quantitative measures of center (median and/or mean) and variability (interquartile range and/or mean absolute deviation), as well as describing any overall pattern and any striking deviations from the overall pattern.
MATH GRADE 7	Use the data collected during your test trials to determine if there is a proportional relationship between the height from the ground (roller coaster starting point) to the speed of your roller coaster.
MATH GRADE 8	Construct and interpret a scatterplot of the speed and height data collected to investigate the relationship between the two quantities.

SHAKE IT UP!

STEAM

SETTING — THE — STAGE

DESIGN CHALLENGE PURPOSE

Design and construct a building prototype that will remain intact during a simulated earthquake.

TEACHER DEVELOPMENT

The earth is made up of four layers: the inner core, the outer core, the mantle, and the crust. The crust and the top part of the mantle make up the surface of the earth. Cracks along the surface of the earth are called **faults**. This is where earthquakes occur. **Earthquakes** are natural disasters consisting of sudden violent shaking or rolling movement of the earth's surface. **Subduction zones** are the areas where tectonic plates collide, forcing one of the plates below the other. This violent collision causes earthquakes. Earthquakes usually do not last long, often less than a minute. However, they can cause immense damage in that short period of time. Humans cannot stop these natural disasters from occurring, but they can work to reduce the impact of these events. A scientist who studies earthquakes is called a **seismologist**.

Note: You will need to build an earthquake shake box prior to the start of the challenge.

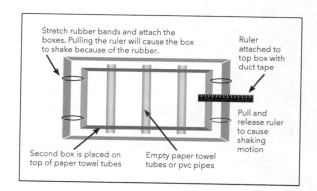

Stretch rubber bands and attach the boxes. Pulling the ruler will cause the box to shake because of the rubber.

Ruler attached to top box with duct tape

Pull and release ruler to cause shaking motion

Second box is placed on top of paper towel tubes

Empty paper towel tubes or pvc pipes

STUDENT DEVELOPMENT

Students need to be familiar with the vocabulary terms related to earthquakes. Students should know that, unlike severe weather or volcanic eruptions, earthquakes are not predictable. The focus for this challenge is on mitigating the effects of earthquakes.

Although this is an engineering challenge, it is closely connected to the science standard that requires students to analyze and interpret data related to natural hazards.

Have students analyze national and international earthquake data. Students should write brief summaries of the data they collected.

Note: Visit the website listed on the inside front cover for information about earthquakes.

STANDARDS

SCIENCE	TECHNOLOGY	ENGINEERING	ARTS	MATH	ELA
MS-ESS3-2	ISTE.1	MS-ETS1-1	Presenting #4	CCSS.MATH. CONTENT.6.SP.B.4	CCSS.ELA-LITERACY. RST.6-8.7
	ISTE.6	MS-ETS1-2	Presenting #5	CCSS.MATH. CONTENT.7.SP.C.6	CCSS.ELA-LITERACY. RST.6-8.9
		MS-ETS1-3	Presenting #6	CCSS.MATH. CONTENT.8.SP.A.4	
		MS-ETS1-4			

SCIENCE & ENGINEERING PRACTICES

Constructing Explanations and Designing Solutions: Apply scientific ideas or principles to design, construct, and/or test a design of an object, tool, process, or system.

CROSSCUTTING CONCEPTS

Cause and Effect: Cause-and-effect relationships may be used to predict phenomena in natural or designed systems.

TARGET VOCABULARY

earthquake

fault line

geology

plate tectonics

seismic activity

stability

structure

subduction zone

MATERIALS

Earthquake Shake Box:
- 2 thick pieces of cardboard (2 large shirt box pieces with corners taped flat works great)
- 2 paper towel tubes (or PVC pipes)
- 2 large rubber bands
- wooden ruler (or similar stick shape), duct tape

Building Materials (one per team):
- 25 straws
- 5 pipe cleaners
- 4 sheets of newspaper
- 5 rubber bands
- string (2 m)
- ¼ cup of glue
- 5 paper clips
- 4 hollow tubes (4 toilet paper tubes)
- 10 craft sticks
- rulers
- scissors

LITERACY CONNECTIONS

Everything Volcanoes and Earthquakes: Earthshaking Photos, Facts, and Fun! by Kathy Furgang

Can You Survive an Earthquake?: An Interactive Survival Adventure by Rachael Hanel

NOTES

DILEMMA ENGAGE

Famous seismologist Dr. Ima Quakin works with the USGS (United States Geological Survey). She, along with her fellow scientists, is working on a plan to minimize the impact of earthquakes in subduction zones. Based on their research, they believe that a major earthquake will occur in the United States within the next few decades. The time to act is now! As Dr. Quakin and the USGS work on their plans for developing early warning systems, developing risk assessment measures, and other technologies that will help mitigate the severity of the earthquake aftermath, she needs engineers to construct more durable buildings that can withstand an earthquake. She needs your help!

MISSION

Construct a prototype of a house that will withstand a simulated earthquake. The prototype must be at least 10 cm but no more than 15 cm in length, width, and height. The building must have a roof. You may use any of the items in the materials list to construct your prototype.

BLUEPRINT EXPLORE

Provide the Individual and Group Blueprint Design Sheets to engineering teams. Have individual students sketch a prototype to present to the other members of their team. Teams will discuss the pros and cons of each sketch and then select one prototype to construct.

ENGINEERING TASK	TEST TRIAL	ANALYZE	REDESIGN
Teams will each design and construct a building prototype that is at least 10 cm but no more than 15 cm in length, width, and height. It must remain intact during a simulated earthquake. *Note:* The teacher should construct the earthquake shake box prior to the start of the challenge. Refer to the diagram on page 62.	One team at a time will test its prototype on the earthquake shake box. The teacher will pull the ruler, which stretches the rubber bands, and then release it to represent an earthquake. The teacher will repeat this 10 times. Repeat tests with remaining teams, testing one team at a time. All other teams observe the testing and the results.	Students must record their observations, explain what happened, and use details to support their explanations.	Students will return to their designs and make changes in colored pencil based on their observations. They must label materials and measurements.

HELPFUL TIPS

- After the Test Trial, have teams take a gallery walk to view other teams' designs for possible ideas to assist them in the Analyze and Redesign portions of the engineering design process.

- If teams are successful on the first try, encourage them to make their prototypes even more efficient. If it is a scenario in which this is not feasible, distribute team members to other teams to be a support for them in making their prototypes more efficient. Alternatively, at teacher discretion, move students on to the Justification portion of the lesson.

- If after the third test the final prototype is still unsuccessful, have students write how they would start over. These challenges are meant to have students build on what they originally designed. If the design proved to be unsuccessful, encourage a reflection or justification on what they would do if they were allowed to start again from scratch.

REFLECTIONS — EXPLAIN & ELABORATE

AFTER TEST TRIAL 1	Did any of the prototypes remain intact and standing after the simulated earthquake? What characteristics did the most successful prototypes have in common?
ANALYSIS	What changes will you make to your prototype to ensure that it remains standing through the simulated earthquake? Explain the reasoning for the changes to your prototype.
AFTER TEST TRIAL 2	Which team's prototype remained intact and standing the longest? What made it more effective than the other prototypes?
ANALYSIS	What changes will you make to your prototype based on your observations? Explain the reasoning for your changes.
AFTER TEST TRIAL 3	Which team of engineers had the most effective prototype? What were the differences between the prototypes?

JUSTIFICATION — EVALUATE

TECHNOLOGY	Use your research on earthquakes to create a computer slideshow that explains the cause of seismic activity and its effects.
MATH GRADE 7	Create a histogram that records the number of major earthquakes in North America over the last 50 years. Determine the range of dates of recorded earthquakes.
MATH GRADE 7	Determine the approximate probability of a tsunami caused by an underwater earthquake occurring in the next decade based on the frequency of occurrence in the last century.
MATH GRADE 7	Construct and interpret a two-way table that summarizes data on the frequency of earthquakes. The first table describes the frequency of occurrence where both of the tectonic plates colliding are continental crusts and the second describes the frequency of occurrence where one of the tectonic plates is an oceanic crust.

WE BUILT A ZOO

3-4 HOURS
TIME FOR COMPLETION

SETTING —THE— STAGE

DESIGN CHALLENGE PURPOSE

Design and construct a 3-D scale model of an enclosure for a zoo animal.

TEACHER DEVELOPMENT

The concept of an appropriately created scale model requires a strong understanding of relative size and distance. This lesson requires students to rely on their previously learned skills of 2-D and 3-D shapes and the construction and deconstruction of those shapes. In addition, the skills of proportional relationships and conversions will be necessary in order to build a scale model of a new animal enclosure at a zoo. The ability to create a scale model will be helpful as students design a new enclosure while keeping in mind the unique needs of their chosen zoo animal.

Note: Prior to the start of the challenge, make four copies of the acreage grid (page 135) and three copies of the rubric (page 134) for each team.

STUDENT DEVELOPMENT

Students must have a working knowledge of shapes, including shape names and angles. They must also know how to use a protractor. Students will need to understand volume, proportions, and scale drawings.

Each team will need to conduct research about its selected animal in order to design an enclosure that meets the needs of the animal.

STANDARDS

SCIENCE	TECHNOLOGY	ENGINEERING	ARTS	MATH	ELA
		MS-ETS1-1	Creating #1	CCSS.MATH.CONTENT.7.G.A.1	
		MS-ETS1-2	Creating #2	CCSS.MATH.CONTENT.7.G.A.2	
		MS-ETS1-3		CCSS.MATH.CONTENT.7.G.A.3	
		MS-ETS1-4		CCSS.MATH.CONTENT.7.G.B.6	
				CCSS.MATH.PRACTICE.MP4	
				CCSS.MATH.PRACTICE.MP5	

SCIENCE & ENGINEERING PRACTICES

Developing and Using Models: Develop or modify a model—based on evidence—to match what happens if a variable or component of a system is changed.

CROSSCUTTING CONCEPTS

Scale, Proportion, and Quantity: Proportional relationships (e.g., speed as the ratio of distance traveled to time taken) among different types of quantities provides information about the magnitude of properties and processes.

Structure and Function: Structures can be designed to serve particular functions by taking into account properties of different materials and how materials can be shaped and used.

TARGET VOCABULARY

circumference

cylinder

diameter

length

pi

proportion

quadrilateral

radius

scale

width

MATERIALS

- construction paper
- scissors
- glue or tape
- building blocks (e.g., snap cubes or plastic interlocking bricks)
- pipe cleaners
- straws
- clear cling film
- rubric (page 134)
- acreage grid (page 135)

LITERACY CONNECTIONS

Architecture: Cool Women Who Design Structures
by Elizabeth Schmermund

Dream Jobs in Math
by Colin Hynson

NOTES

STEAM —IN— ACTION

DILEMMA ENGAGE

Ms. Annie Mall has worked at the local zoo for the past five years. During her time there, she has noticed that there is a lot of space that is not being utilized to its full potential. To correct this situation, Annie shared her concerns with the president of the zoo. The zoo president was excited to hear that Annie was watching out for the animals and asked her to find a group of community members to help her redesign her favorite animal's enclosure.

Can you help Annie by creating a 3-D scale model of a zoo animal enclosure?

MISSION

Create a 3-D scale model zoo enclosure.

Enclosure Requirements:

1. Takes up no more than 2 acres of land but must use 75% of those 2 acres.

2. At least 25% of the enclosure must be reserved for shelter for the animal.

3. Must include at least five features from the animal's natural habitat.

4. Must have at least two cylindrical structures in the enclosure and a circular watering hole.

5. All structures must be labeled and include the amount of land used by each structure.

BLUEPRINT EXPLORE

Provide the Individual and Group Blueprint Design Sheets to engineering teams. Have individual students sketch a prototype to present to the other members of their team. Team members will discuss the pros and cons of each sketch and then select one prototype to construct.

ENGINEERING TASK

Each team will select an animal to build an enclosure for. Then teams will each create a 3-D scale model of the enclosure.

Note: Explain that each centimeter in length will be equal to 8 ft.

Encourage teams to consider the size of their animal and how that will affect the amount of space it needs.

TEST TRIAL

Teams will use the rubric to analyze another team's model to ensure that it meets all requirements.

Note: Teams will construct the 3-D model during Test Trial 3.

ANALYZE

Teams will reflect on the feedback from the rubric.

Teams should be allowed to observe the other designs to gather ideas and reflect in order to improve their prototypes.

REDESIGN

Teams will utilize the feedback to make adjustments to improve their zoo enclosure.

HELPFUL TIPS

- After the Test Trial, have teams take a gallery walk to view other teams' designs for possible ideas to assist them in the Analyze and Redesign portions of the engineering design process.

- If teams are successful on the first try, encourage them to make their prototypes even more efficient. If it is a scenario in which this is not feasible, distribute team members to other teams to be a support for them in making their prototypes more efficient. Alternatively, at teacher discretion, move students on to the Justification portion of the lesson.

- If after the third test the final prototype is still unsuccessful, have students write how they would start over. These challenges are meant to have students build on what they originally designed. If the design proved to be unsuccessful, encourage a reflection or justification on what they would do if they were allowed to start again from scratch.

REFLECTIONS — EXPLAIN & ELABORATE

AFTER TEST TRIAL 1	What rubric score did your prototype receive after Test Trial 1? Was there an area on the rubric where your prototype did not receive full points?
ANALYSIS	How will you improve your prototype?
AFTER TEST TRIAL 2	What rubric score did your prototype receive after Test Trial 2?
ANALYSIS	Did your total rubric score improve? Are there any other ways you can improve your prototype? Explain.
AFTER TEST TRIAL 3	What was your final rubric score? Do you think your chosen animal would enjoy living in the enclosure you designed? Given the opportunity to redesign your prototype, what would you change and why?

JUSTIFICATION — EVALUATE

ELA/ TECHNOLOGY	Write a public service announcement (PSA) that raises awareness of the importance of providing a natural habitat for zoo animals. Present your PSA as a newscast or record a video.
ARTS	Create a 2-D sketch of your enclosure prototype. Include the chosen animal in the sketch.
MATH	Calculate the area of each of the noncircular structures inside your animal enclosure. Calculate the volume of the cylindrical structures inside your animal enclosure.

LAST CRITTER STANDING

SETTING THE STAGE

DESIGN CHALLENGE PURPOSE

Design a new animal species that will flourish in the ecosystem of a newly discovered planet.

TEACHER DEVELOPMENT

An **ecosystem** is a community of plants and animals that live together and interact with each other as well as with the nonliving elements in the environment. Animals or plants with similar needs may compete with each other for the limited resources, such as food, water, and space, in their ecosystem. This competition can affect a plant or animal's growth, reproduction, and survival.

While animals compete for the resources in an area, they are also dependent on each other. The balance of an ecosystem is what keeps it healthy and its population equally distributed. When something happens to the ecosystem, such as drought, flood, or fire, it affects the animals and plants that live there.

STUDENT DEVELOPMENT

Students will need to have an understanding of ecosystems and the interdependence of animals and plants living within an ecosystem.

Review ecosystems with your students and remind them that ecosystems are part of a bigger environment called a biome. A **biome** is a large natural community made up of living and nonliving things that interact with each other. Ecosystems exist within a biome.

Lesson Idea: Have students create a foldable to place inside their science notebooks. Fold a large piece of paper into thirds. Glue one folded third to a page inside the science notebook. Use the remaining five sections (two sections on one side, three on the other) to record information about five major biomes. Students should include the types of animals and plants that live there as well as the biomes' climate.

Note: Visit the website listed on the inside front cover for more information about biomes.

STANDARDS

SCIENCE	TECHNOLOGY	ENGINEERING	ARTS	MATH	ELA
MS-ESS3-2	ISTE.1		Creating #1		CCSS.ELA-LITERACY.WHST.6-8.7
	ISTE.3		Creating #2		CCSS.ELA-LITERACY.WHST.6-8.4
			Creating #3		

SCIENCE & ENGINEERING PRACTICES

Engaging in Argument from Evidence: Construct, use, and/or present an oral and written argument supported by empirical evidence and scientific reasoning to support or refute an explanation or a model for a phenomenon or a solution to a problem.

Evaluate competing design solutions based on jointly developed and agreed-upon design criteria.

CROSSCUTTING CONCEPTS

Cause and Effect: Cause-and-effect relationships may be used to predict phenomena in natural or designed systems.

TARGET VOCABULARY

biome

carnivore

diurnal

ecosystem

food chain

herbivore

omnivore

nocturnal

predator

rain forest

species

vegetation

MATERIALS

- blank paper for designing and labeling animal
- resource cards (page 136)
- challenge cards (page 136-137)

Note: Print both the resource and challenge cards on card stock or laminate for reuse.

Print three copies of the resource cards. There are normally only three test trials, but in this challenge there are six challenge card scenarios. This allows teams to either pick the three they prefer or complete additional test trials.

LITERACY CONNECTIONS

DK Eyewitness Books: The Amazon by DK

What's Up in the Amazon Rainforest by Ginjer L. Clarke

NOTES

STEAM IN ACTION

DILEMMA ENGAGE

It is the year 3010, and Earth is overcrowded! Citizens have started to colonize planets beyond their solar system. The Interstellar Planetary Exploration Corporation (IPEC) has found a planet they have named Amazon, after the rain forest on Earth. The climate and plant life on planet Amazon is very similar to the forest biomes on Earth. Unfortunately, a recent plague has swept through the planet, killing many of the native animal species. The plague originated from the animals the Interstellar Planetary Exploration Corporation brought from Earth. After this huge disaster, the chairman of IPEC, Mr. D. Wild, needs his engineers to design a brand-new species of animal to help repopulate part of the rain forest ecosystem on planet Amazon.

MISSION

Design an animal that can survive in the rain forest on planet Amazon. Label the animal's characteristics, coloring, and measurements. Include whether the animal is an omnivore, carnivore, or herbivore, and describe where it finds its food.

BLUEPRINT EXPLORE

Provide the Individual and Group Blueprint Design Sheets to engineering teams. Have individual students sketch a prototype to present to the other members of their team. Team members will discuss the pros and cons of each sketch and then select one prototype to construct.

Remind students to label the animal's characteristics, coloring, and measurements. Students should include whether the animal is an omnivore, carnivore, or herbivore. They should describe what it eats and where it finds its food.

ENGINEERING TASK	TEST TRIAL	ANALYZE	REDESIGN
Teams will each design an animal prototype that will compete for food and other resources. During Test Trial 1, each team will present its animal prototype to another team, describing and defending the animal's ability to survive in the habitat on planet Amazon. After the first test trial, teams will present any changes they made to their animal before finalizing their animal designs and proceeding to the second phase of the challenge.	Each team takes five resource cards. The cards represent resources the animal uses to survive. The teacher selects one challenge card. This card describes the teams' next steps. Animal prototypes survive by adding up points on the resource cards they did not have to forfeit. Resource cards are placed back in the deck and reshuffled for the next test trial. If a team does not have the needed resource cards or if its animal prototype doesn't have the characteristics mentioned on the challenge card, the team forfeits all of its resource cards and receives no points for that round.	Teams will analyze the feedback they received after presenting their animal to another team and make any necessary changes. After the second test trial, teams can make changes to help them achieve a higher score in the next test trial. *Note:* Teams should determine what traits or adaptations they need to label or may wish to add.	Students will return to their designs and make changes in colored pencil based on their observations during the test trial.

HELPFUL TIPS

- Teams should keep track of their points for each round/test trial and calculate the final total after the third test trial.

- The resource card points are intentionally random. This allows some elements of chance and keeps students engaged.

STEAM Design Challenges Gr. 6–8 © 2018 Creative Teaching Press

REFLECTIONS — EXPLAIN & ELABORATE

AFTER TEST TRIAL 1	What characteristics did the animal prototypes with the highest scores have in common? Explain your observations.
ANALYSIS	What changes will your team make to your prototype? Explain why you will make these changes.
AFTER TEST TRIAL 2	Did your animal survive the second challenge? Which team's prototype has the most points? Explain why that prototype has the most points so far.
ANALYSIS	What changes will you make to your prototype? Explain why you will make these changes.
AFTER TEST TRIAL 3	What animal received the most points? Explain why you think it received the most points during the trials.

JUSTIFICATION — EVALUATE

TECHNOLOGY	Research an ecosystem and the impact humans have on it. Then create a computer slideshow to present your research.
ELA	Generate a question related to ecosystems, resources, and plant and animal life. Then research and write a short report that answers this question.
ARTS	Create a 3-D model of your animal using clay, cardboard, or other available materials.

EXTRATERRESTRIAL GARDENING

STEAM

1-3 HOURS
TIME FOR COMPLETION

SETTING THE STAGE

DESIGN CHALLENGE PURPOSE

Use your knowledge of photosynthesis to design a way to grow plants in a space substation on Mars.

TEACHER DEVELOPMENT

Photosynthesis is the process that plants use to take **carbon dioxide**, water, and sunlight and chemically convert them into carbohydrates (three-carbon sugars) and oxygen. In addition to knowledge of photosynthesis, students will also need to understand (1) the two stages that occur inside the **chloroplasts**; (2) the **light-dependent reactions** that produce ATP (a small molecule that transports chemical energy within cells), NADPH (a chemical compound used to help turn carbon dioxide into glucose), and oxygen; and (3) the Calvin Cycle (the set of chemical reactions that take place in chloroplasts during photosynthesis and take place after the energy has been captured from sunlight), which produces three-carbon sugars (glucose).

Even though it is important to understand the role that **cellular respiration** plays during photosynthesis, this lesson focuses on photosynthesis.

There is a scene in the movie *The Martian* that shows how a botanist constructs a system to grow plants on Mars. This scene may provide students with ideas about how photosynthesis plays a part in helping a greenhouse grow plants.

Note: Visit the website listed on the inside front cover for more information about photosynthesis and for a clip from the movie *The Martian*.

STUDENT DEVELOPMENT

In order to understand how to solve the problems presented in this challenge, students need to have a good understanding of the process of **photosynthesis**. They need to know the **chemical equation** for photosynthesis:

$$6CO_2 + 6H_2O \xrightarrow{energy} C_6H_{12}O_6 + 6O_2$$

(where sunlight is needed).

Students need to know what each component stands for: CO_2 = carbon dioxide, H_2O = water, $C_6H_{12}O_6$ = glucose, and O_2 = oxygen

Students will also need time to research the atmosphere and gases on Mars.

Note: Visit the website listed on the inside front cover for more information about Mars.

STANDARDS

SCIENCE	TECHNOLOGY	ENGINEERING	ARTS	MATH	ELA
MS-LS1-6	ISTE.3	MS-ETS1-1	Creating #1	CCSS.MATH.PRACTICE.MP2	CCSS.ELA-LITERACY.W.6.1
MS-LS1-7	ISTE.6	MS-ETS1-2	Creating #2	CCSS.MATH.PRACTICE.MP4	CCSS.ELA-LITERACY.W.7.1
		MS-ETS1-3	Creating #3		CCSS.ELA-LITERACY.W.8.1
		MS-ETS1-4			

SCIENCE & ENGINEERING PRACTICES

Constructing Explanations and Designing Solutions: Construct a scientific explanation based on valid and reliable evidence obtained from sources (including the students' own experiments) and the assumption that theories and laws that describe the natural world operate today as they did in the past and will continue to do so in the future.

CROSSCUTTING CONCEPTS

Energy and Matter: Within a natural system, the transfer of energy drives the motion and/or cycling of matter.

TARGET VOCABULARY

balanced equation

carbon dioxide

cellular respiration

chemical equation

chemical reaction

condensation

energy

evaporation

light-dependent reactions

loam

matter

perspiration

photosynthesis

precipitation

sugar

transport

water cycle

MATERIALS

- aluminum foil
- clay
- small container (to simulate an oxygen tank)
- paper (for labeling)
- paper cups
- plastic wrap
- craft sticks
- rulers
- seedlings (sprouted beans)
- string
- tape
- water (1 cup per team)
- rubric (page 138)

LITERACY CONNECTIONS

"The Martian becomes reality: Four crops found edible when grown on simulated Mars soil."
by Wageningen University

"Can we safely eat plants grown on Mars?"
by Wageningen University

"Food for Mars: Green beans and potatoes"
by Natasha Schön

NOTES

82

STEAM
— IN —
ACTION

DILEMMA · ENGAGE

It is the year 2035 and there are five people living inside a substation on Mars. There is a problem! A rocket carrying supplies from Earth to the substation was canceled, and another shipment of food from Earth is not due to arrive for another year. In order to survive, the people on the substation need to grow plants for food. They have seedlings, some water, some nitrogen-rich loam, plastic sheeting, and aluminum insulation material that can be used to help them grow plants for food. Can you help the substation inhabitants design a way to grow and sustain plants?

MISSION

Build a model greenhouse that would allow plants to grow on Mars.

Requirements:

1. A building plan that contains:
   a. The equation for photosynthesis and describes how it is created and sustained within a greenhouse.
   b. The equation for the chemical composition of Mars' atmosphere and connects it to the process of photosynthesis, recording energy change and transfer. The equation must be balanced.

2. Labels on the model showing where the different stages of photosynthesis take place.

BLUEPRINT · EXPLORE

Provide the Individual and Group Blueprint Design Sheets to engineering teams. Have individual students sketch a prototype to present to the other members of their team. Team members will discuss the pros and cons of each sketch and then select one prototype to construct.

ENGINEERING TASK	TEST TRIAL	ANALYZE	REDESIGN
Teams will each first create a detailed plan, labeling the exchange of CO_2 and O_2 that they will demonstrate with their prototypes. Once the plan is approved, each team will create a model that provides optimal conditions (including all labels from the plan) for growing plants on Mars. *Note:* The labels must include chemical names, show the flow of energy from the matter used, and demonstrate an understanding of the chemical eqution for photosynthesis.	Teams will each design a detailed plan with labels written to show the exchange of CO_2 and O_2 within the planned prototype. Once approved, students need to show the same balanced equation within their prototype and be able to explain how it would work. The prototype will have the labeled equations.	Teams should check their math to ensure that the model growing habitat would provide the needed elements and molecules for photosynthesis to occur. Teams should be allowed to observe other teams' designs to gather ideas and reflect in order to create a more efficient prototype.	If the team equation doesn't work due to missing information, teams should add the missing information. They should label and explain their changes on their prototypes.

HELPFUL TIPS

- After the Test Trial, have teams take a gallery walk to view other teams' designs for possible ideas to assist them in the Analyze and Redesign portions of the engineering design process.

- If teams are successful on the first try, encourage them to make their prototypes even more efficient. If it is a scenario in which this is not feasible, distribute team members to other teams to be a support for them in making their prototypes more efficient. Alternatively, at teacher discretion, move students on to the Justification portion of the lesson.

- If after the third test the final prototype is still unsuccessful, have students write how they would start over. These challenges are meant to have students build on what they originally designed. If the design proved to be unsuccessful, encourage a reflection or justification on what they would do if they were allowed to start again from scratch.

REFLECTIONS — EXPLAIN & ELABORATE

AFTER TEST TRIAL 1	Did your prototype have all of the requirements to sustain plants on Mars?
ANALYSIS	What part of your design should be changed in order to show the correct chemical equations?
AFTER TEST TRIAL 2	Did your prototype meet all of the requirements of the rubric?
ANALYSIS	What can you add, delete, or change in order to meet all of the rubric requirements?
AFTER TEST TRIAL 3	Did your plan, prototype, and explanation meet the challenge requirements?

JUSTIFICATION — EVALUATE

SCIENCE	Demonstrate how cellular respiration would work with your prototype and plan for the growing habitat.
TECHNOLOGY	Create a slideshow presentation that invites people to apply for a farming position on Mars. They will use what they learned from this challenge to present the requirements for a farming habitat.
ELA	Write a newsletter in support of a crowdfunding campaign to complete further research on the growth of plants on a Martian substation. Use evidence from your research and draw on your experience to argue the merits of farming on Mars.
ARTS	Draw the finished and flourishing Martian substation greenhouse that your team designed. It should include a text box and/or diagram with labels to explain how photosynthesis would occur for the plants using the Martian atmosphere and the supplies you used to construct it.

LIVING CELLS

STEAm

SETTING
— THE —
STAGE

DESIGN CHALLENGE PURPOSE

Create a model of a plant or animal cell that demonstrates the functions of the various organelles within the cell.

TEACHER DEVELOPMENT

This challenge focuses on cells and their structure. The cell has a very specific function, and each of the components of that cell work together to ensure that it functions properly. The components of a cell are the **cell membrane**, **cell wall** (plant cells only), **cytoplasm**, and various **organelles**, which include the **nucleus**, **mitochondria**, **vacuoles**, and **chloroplast** (plant cells only).

This challenge will provide the opportunity for students to dive deeper into the organelles and how they help the cell function as one organism.

Note: Visit the website listed on the inside front cover for more information about cells.

STUDENT DEVELOPMENT

Students will gather information about their team's organelle and how it contributes to the living cell. The function within the cell will be necessary for the completion of this challenge. The teacher should make sure that all organelles are represented in this challenge.

STANDARDS

SCIENCE	TECHNOLOGY	ENGINEERING	ARTS	MATH	ELA
MS-LS1-1	ISTE.3	MS-ETS1-1	Creating #2		CCSS.ELA-LITERACY.RST.6-8.2
MS-LS1-2		MS-ETS1-2	Presenting #6		
		MS-ETS1-3			
		MS-ETS1-4			

SCIENCE & ENGINEERING PRACTICES

Developing and Using Models: Develop and/or revise a model to show the relationships among variables, including those that are not observable but predict observable phenomena.

CROSSCUTTING CONCEPTS

Structure and Function: Complex and microscopic structures and systems can be visualized, modeled, and used to describe how their function depends on the shapes, composition, and relationships among its parts, therefore complex natural and designed structures/systems can be analyzed to determine how they function.

TARGET VOCABULARY

cell membrane

cell wall

chloroplast

cytoplasm

mitochondria

nucleus

vacuoles

MATERIALS

- butcher paper or construction paper
- colored pencils or markers
- fabric
- pipe cleaners
- cereal boxes
- straws
- modeling clay
- other recyclable materials
- rubric (page 139)

LITERACY CONNECTIONS

Cell Wars: In the Beginning by A. Miles

NOTES

STEAM —IN— ACTION

DILEMMA | ENGAGE

Dr. Justin Cellular is in charge of the middle school exhibits at the Marvelous Museum of Science. Dr. Cellular has had the same exhibits in his museum for the last 10 years and is ready for something new and exciting. Dr. Cellular wants one of the new exhibits to be about living cells. He is looking to the local middle schools to help him create an exhibit that informs visitors about how a living cell works. Can your team design a model to help Dr. Cellular?

MISSION

Create a model that highlights the functions of a living cell.

BLUEPRINT | EXPLORE

Provide the Individual and Group Blueprint Design Sheets to engineering teams. Have individual students sketch a prototype to present to the other members of their team. Team members will discuss the pros and cons of each sketch and then select one prototype to construct.

ENGINEERING TASK

 TEST TRIAL

 ANALYZE

 REDESIGN

ENGINEERING TASK	TEST TRIAL	ANALYZE	REDESIGN
After conducting research on the organelle of their choice, teams will each construct a model of that organelle and then combine it with the other teams' model organelles to create a class model of a living cell. *Note:* Teachers can divide the teams into a plant cell group and an animal cell group. The class model can be done twice, one time for each type of cell.	The teacher will act as an "intruder" into the cell. Teams must demonstrate and explain how their organelle will respond to the intruder. *Note:* The combined organelles must realistically respond to your intrusion into the larger classroom living cell model. You will determine if students understand the function of each organelle in the cell.	Teams will use the rubric to analyze their organelle's function within the larger living cell.	Each team will modify their organelle as needed before testing again.

HELPFUL TIPS

- After the Test Trial, have teams take a gallery walk to view other teams' designs for possible ideas to assist them in the Analyze and Redesign portions of the engineering design process.

- If teams are successful on the first try, encourage them to make their prototypes even more efficient. If it is a scenario in which this is not feasible, distribute team members to other teams to be a support for them in making their prototypes more efficient. Alternatively, at teacher discretion, move students on to the Justification portion of the lesson.

- If after the third test the final prototype is still unsuccessful, have students write how they would start over. These challenges are meant to have students build on what they originally designed. If the design proved to be unsuccessful, encourage a reflection or justification on what they would do if they were allowed to start again from scratch.

REFLECTIONS — EXPLAIN & ELABORATE

AFTER TEST TRIAL 1	Did your organelle react appropriately when the "intruder" interacted with it inside the larger living cell?
ANALYSIS	What changes will you make to your organelle model?
AFTER TEST TRIAL 2	Did your organelle react appropriately to the intruder? How did it respond to the intruder?
ANALYSIS	What changes do you need to make to your organelle to better meet the needs of the cell?
AFTER TEST TRIAL 3	Reflect on the changes you made to your model. Then decide what was the most important change that you made to your organelle. Explain.

JUSTIFICATION — EVALUATE

TECHNOLOGY	Use a microscope to view various slides of cells. Then sketch and label what you see.
ELA	Write a letter to Dr. Cellular explaining why your organelle for the model of the living cell should be chosen as the next great exhibit at the museum.
ARTS	Create a 3-D model of an animal or plant cell with all organelles labeled.

LIVE ON THE SCENE

STeAm

SETTING
—THE—
STAGE

DESIGN CHALLENGE PURPOSE

Write and perform a news report that focuses on an organism in an ecosystem and highlights how energy flows through that ecosystem.

TEACHER DEVELOPMENT

This challenge asks students to model how energy flows (energy web) through a selected ecosystem. The students will need to have background knowledge on both living and nonliving organisms that exist in that ecosystem.

When looking at an energy web, the students must know that the flow of energy is shown by an arrow going from one thing to another. A common misconception is that the larger organism has the arrow pointing to the smaller organism because it is consuming it for energy. This is incorrect, and students will need to know that the smaller organism is providing energy to the larger organism. Therefore the arrow must go from the organism providing the energy to the organism that is receiving that energy.

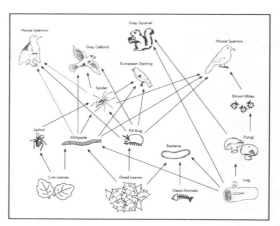

Note: Visit the website listed on the inside front cover for more information about energy webs.

STUDENT DEVELOPMENT

Students will need to accurately demonstrate in written form their understanding of how energy flows from one organism to another.

STANDARDS

SCIENCE	TECHNOLOGY	ENGINEERING	ARTS	MATH	ELA
MS-LS2-3	ISTE.3		Creating #1		CCSS.ELA-LITERACY.W.6.3
			Creating #2		CCSS.ELA-LITERACY.W.6.4
			Creating #3		CCSS.ELA-LITERACY.W.6.5
			Performing #4		CCSS.ELA-LITERACY.SL.6.1
			Performing #5		
			Performing #6		

SCIENCE & ENGINEERING PRACTICES

Developing and Using Models: Develop and/or use a model to predict and/or describe a phenomena.

CROSSCUTTING CONCEPTS

Energy and Matter: Within a natural system, the transfer of energy drives the motion and/or cycling of matter.

The transfer of energy can be tracked as energy flows through a natural system.

TARGET VOCABULARY

arctic

desert

ecosystem

flow of energy

matter

nonliving

ocean

plant

river

tundra

MATERIALS

- paper
- research materials
- electronic presentation program such as PowerPoint
- rubric (page 140)

LITERACY CONNECTIONS

One Day in the Woods by Jean Craighead George

One Day in the Tropical Rain Forest by Jean Craighead George

One Day in the Desert by Jean Craighead George

NOTES

STEAM
—IN—
ACTION

DILEMMA   ENGAGE

While on an assignment in the jungle, field scientist Jessica Journey investigated this ecosystem and all of its inhabitants. She needs to report on the information she gathered about the plants, animals, and insects living in this ecosystem, but she needs your help. Can your team help Jessica Journey by writing and performing a news report about what she saw?

Note: You are not limited to the jungle ecosystem. You may change it to best fit your current course of study.

MISSION

Write and present a news report from the perspective of an organism in the jungle ecosystem. The report must include how the organism contributes to the cycling of matter and how it provides or receives energy within the system. The report must also describe at least 10 animals, 5 plants, and 3 nonliving parts of the ecosystem.

Once the news report is written, your team will perform it in front of an audience.

BLUEPRINT   EXPLORE

Provide the Individual and Group Blueprint Design Sheets to engineering teams. Have individual students list three to four scenarios they may want to explore in their news report and then present those ideas to the other members of their team. Team members will discuss the pros and cons of each list and then select one scenario they will use for the challenge.

ENGINEERING TASK	TEST TRIAL	ANALYZE	REDESIGN
Each team will need to research an ecosystem and write a detailed news report from the perspective of an organism in that ecosystem. The team will explain the role of the organism in that ecosystem; how energy flows in the ecosystem; and how matter is cycled throughout the ecosystem.	Teams will use the rubric to assess the news report performed by the other teams.	Teams will use the rubric feedback to make changes to their reports.	Teams will have the opportunity to perform their news reports again.

HELPFUL TIPS

- After the Test Trial, have teams take a gallery walk to view other teams' presentations for possible ideas to assist them in the Analyze and Redesign portions of the engineering design process.

- If teams receive a perfect score on the first try, encourage them to make their presentations even more detailed and engaging. If it is a scenario in which this is not feasible, distribute team members to other teams to be a support for them in making their presentation more accurate. Alternatively, at teacher discretion, move students on to the Justification portion of the lesson.

- If the final presentation is still inaccurate, have students write how they would start over. These challenges are meant to have students build on what they originally presented. If the presentation proved to be unsuccessful, encourage a reflection or justification on what they would do if they were allowed to start again from scratch.

REFLECTIONS — EXPLAIN & ELABORATE

AFTER TEST TRIAL 1	What was your team rubric score?
ANALYSIS	What will you change in order to increase your overall rubric score?
AFTER TEST TRIAL 2	Did your rubric score improve? Are there any other areas where you can improve?
ANALYSIS	How might you make your presentation more engaging?
AFTER TEST TRIAL 3	If you had the opportunity to transport yourself to a different ecosystem, would you and your teammates do it? Why or why not?

JUSTIFICATION — EVALUATE

TECHNOLOGY	Record your news report, and post it on the class website.
ELA	Select another organism from a different ecosystem, and write a new adventure from that organism's perspective. *Note:* If you chose a consumer for your first adventure, select a producer this time. If you were a living thing in the first adventure, select a nonliving thing this time.
ARTS	Create a 3-D representation of your ecosystem. Label all of the parts.

MAGICALLY MOVE IT!

STEAM

1 HOUR

TIME FOR COMPLETION

SETTING —THE— STAGE

DESIGN CHALLENGE PURPOSE

Construct a paper glider that travels a distance of 100 feet and lands on a designated target.

TEACHER DEVELOPMENT

Gliders can stay in the air a long time due to pockets of air called **updrafts**. When the glider gains altitude from these updrafts, potential energy is increased. **Potential energy** is the amount of **stored** **energy** an object has due to its **position**. You can create **lift** by moving a poster board behind a paper airplane in order to keep it in the air.

STUDENT DEVELOPMENT

Gliders can stay in the air a long time because of pockets of air called **updrafts**. When the glider gains altitude from these updrafts, potential energy is increased. **Potential energy** is the amount of **stored energy** an object has due to its **position**. **Lift** is a force that is created by differences in air pressure. Lift affects the wings of a plane. According to NASA, lift is due to Newton's third law because a solid is affected by the flow of a gas or liquid and lift acts perpendicular to the motion. A wing on a plane is affected by the motion of the gas flow below and above the wing. **Drag** opposes lift.

Note: Visit the website listed on the inside front cover for more information about gliders.

STANDARDS

SCIENCE	TECHNOLOGY	ENGINEERING	ARTS	MATH	ELA
MS-PS3-2	ISTE.2	MS-ETS1-1	Creating #1	CCSS.MATH. CONTENT.6.SP.B.4	CCSS.ELA-LITERACY. RST.6-8.7
MS-PS3-5		MS-ETS1-2		CCSS.MATH. CONTENT.6.SP.B.5	CCSS.ELA-LITERACY. RST.6-8.3
		MS-ETS1-3			
		MS-ETS1-4			

SCIENCE & ENGINEERING PRACTICES

Developing and using Models: Develop a model to describe unobservable mechanisms.

Engaging in Argument from Evidence: Construct, use, and/or present an oral and written argument supported by empirical evidence and scientific reasoning to support or refute an explanation or a model for a phenomenon or a solution to a problem.

CROSSCUTTING CONCEPTS

Systems and System Models: Models can be used to represent systems and their interactions—such as inputs, processes, and outputs—and energy and matter flows within systems.

Energy and Matter: Energy may take different forms (e.g., energy in fields, thermal energy, and energy of motion).

TARGET VOCABULARY

altitude

drag

forces at a distance

glider

kinetic energy

lift

position

potential energy

stored energy

updraft

MATERIALS

- white copy paper
- poster board
- ruler
- tape
- tape measure
- masking tape
- scissors

LITERACY CONNECTIONS

Planes, Gliders, and Paper Rockets: Simple Flying Things Anyone Can Make—Kites and Copters, Too!
By Rick Schertle and James Floyd Kelly

NOTES

STEAM
—IN—
ACTION

DILEMMA ENGAGE

The famous movie director Wynn A. Oscar has written a script for a movie about gliders. Now he's trying find a producer to help him make his movie. The producer I. M. Loaded is interested, but she doesn't know much about gliders. She wants to see a model glider before she agrees to help with the movie. Mrs. Loaded wants the director to simulate the flight of a glider using a very specific list of items. Mr. Oscar may only use some paper, a poster board, a ruler, tape, and scissors. Mrs. Loaded has challenged the director to make the glider travel 100 feet without touching it and then land it on a specific target. Can you help Mr. Oscar make a glider that can meet these requirements?

MISSION

Construct a paper glider that can travel a distance of 100 feet without being touched. Then land the glider on a designated target.

BLUEPRINT EXPLORE

Provide the Individual and Group Blueprint Design Sheets to engineering teams. Have individual students sketch a glider prototype and develop a strategy to keep it in the air. Have them then present their ideas to the other members of their team. Teams will discuss the pros and cons of each sketch and then select one prototype to construct.

ENGINEERING TASK	TEST TRIAL	ANALYZE	REDESIGN
Each team will construct a paper glider and develop a strategy for keeping the glider in the air long enough to land it on a designated target.	Teams will test their gliders and strategies for creating the updraft that provides the lift for their gliders. Teams should measure and record the distance their gliders travel from the starting point. *Note:* This challenge requires a large testing space, such as a long hallway or open gymnasium. Use masking tape to mark the starting line and the ending point/target 100 feet from the starting line. It is suggested that all teams observe each other test their gliders and record the distance each glider travels.	Teams will analyze the data collected during testing and discuss the features and techniques that made a flight successful. Teams should create a graphic display of their collected data (e.g., dot plots, coordinate grid graphs, box plots, or histograms).	Teams should discuss the different glider designs and techniques and how they might be improved to increase the glider's travel distance. Changes to the original design should be made with a colored pencil on the Blueprint Design Sheet.

HELPFUL TIPS

- After the Test Trial, have teams take a gallery walk to view other teams' designs for possible ideas to assist them in the Analyze and Redesign portions of the engineering design process.

- If teams are successful on the first try, encourage them to make their prototypes even more efficient. If it is a scenario in which this is not feasible, distribute team members to other teams to be a support for them in making their prototypes more efficient. Alternatively, at teacher discretion, move students on to the Justification portion of the lesson.

- If after the third test the final prototype is still unsuccessful, have students write how they would start over. These challenges are meant to have students build on what they originally designed. If the design proved to be unsuccessful, encourage a reflection or justification on what they would do if they were allowed to start again from scratch.

STEAM Design Challenges Gr. 6–8 © 2018 Creative Teaching Press

REFLECTIONS   EXPLAIN & ELABORATE

AFTER TEST TRIAL 1	Teams should measure and record the distance their glider traveled from the starting point. Then record the distance the other teams' gliders traveled.
ANALYSIS	What changes will your team make to your glider? Why? What features of your glider and technique contributed to the failure or success of your glider?
AFTER TEST TRIAL 2	Did any team land on the designated target?
ANALYSIS	Explain why you think the successful team's design made it possible for the glider to land on the target.
AFTER TEST TRIAL 3	Did any team improve the distance its glider traveled? What changes did your team make to your glider and poster board technique that made it possible for your glider to travel its greatest distance?

JUSTIFICATION   EVALUATE

ELA/ TECHNOLOGY	Create a video commercial advertising your glider's design and the technique you used to make it move.
ARTS	Draw and label a diagram showing the forces that acted on the glider and how the poster board technique your team used played a role in those forces being produced.

MODIFY THE ROOM

STEAM

SETTING THE STAGE

DESIGN CHALLENGE PURPOSE

Create an aesthetically pleasing barrier that muffles the sound of music coming from a room.

TEACHER DEVELOPMENT

Sound waves can be measured in decibels (dB). A zero is the measurement of the softest sound you can hear. A whisper is 20 dB, a washing machine is 75 dB, a rock concert is 110 dB, a jet plane is 130 dB, and a rocket launch is above 165 dB. The normal decibel level that is safe to listen to for an extended amount of time is 85 dB.

The teacher will need to set the volume of a device that will be used during the challenge to a level of 110 dB. This will be the standard for which all of the test trials will occur. The device must be small enough to fit inside the teams' prototype rooms. An iPod or other similarly sized device is ideal.

Note: Visit the website listed on the inside front cover for more information about sound.

STUDENT DEVELOPMENT

Sound is measured in decibels and is actually energy that travels in waves. Sound can be measured in two ways: frequency and amplitude. Frequency is the number of sound vibrations in one second and amplitude is how powerful the waves are. The power of the waves is measured in decibels (dB). The average decibel level that is safe to listen to for an extended amount of time is 85 dB. A rock concert measures about 110 dB, which would be similar to a dance club.

To dampen sound waves, the vibrations in the waves are dissipated before they build. This is done through an approach called soundproofing. Soundproofing is a combination of different ways to reduce sound. Some examples of soundproofing include increasing the distance between the origin of the sound and the people listening to the sound, trapping the sound waves (absorption), and scattering the sound waves (diffusion).

STANDARDS

SCIENCE	TECHNOLOGY	ENGINEERING	ARTS	MATH	ELA
MS-PS4-2		MS-ETS1-1	Creating #1	CCSS.MATH. CONTENT.6.SP.B.4	
		MS-ETS1-2			
		MS-ETS1-3			
		MS-ETS1-4			

SCIENCE & ENGINEERING PRACTICES

Developing and Using Models: Develop and/or use a model to predict and/or describe phenomena.

CROSSCUTTING CONCEPTS

Structure and Function: Structures can be designed to serve particular functions by taking into account properties of different materials and how materials can be shaped and used.

TARGET VOCABULARY

absorb

decibel

reflection

refraction

sound wave

transmit

MATERIALS

- cardboard
- Styrofoam
- tape
- newspaper
- plastic sheets (the thickness used for photo frames works well)
- scissors
- tape
- sound decibel smart phone app or sound level meter

LITERACY CONNECTIONS

Light and Sound: Energy, Waves, and Motion
by Barbara R. Sandall, EdD and LaVerne Logan

NOTES

STEAM — IN — ACTION

DILEMMA | ENGAGE

Mr. Edward Cool Beans is building a new dance club and restaurant. He wants to play loud music in the dance club, but he is worried that the sound of the bass may be too loud for the people eating in the restaurant. He needs your help to design a visually appealing wall that will separate the dance club from the restaurant. The wall needs to muffle the sounds of the music coming from the dance club so that the diners in the restaurant can enjoy their meal.

MISSION

Create a visually appealing wall that also reduces noise coming from one side.

BLUEPRINT | EXPLORE

Provide the Individual and Group Blueprint Design Sheets to engineering teams. Have individual students sketch a prototype to present to the other members of their team. Team members will discuss the pros and cons of each sketch and then select one prototype to construct.

ENGINEERING TASK	TEST TRIAL	ANALYZE	REDESIGN
Teams will each construct a room with a visually appealing wall prototype that will muffle the sound coming from one side of the wall.	Teams will test their wall prototypes using the sound decibel app and the device playing music that was preset by the teacher at 110 dB. Teams will place the smart phone with the sound decibel app on one side of the wall of the room they constructed and place the music device on the other side of the wall. Teams will record the decibel reading.	Teams will review the data they collected and compare it to the other teams' data. Teams should determine what improvements to make to their walls in order to decrease the decibel reading. Teams should be allowed to observe the other teams' designs to gather ideas and reflect in order to create a more efficient prototype.	Teams can redesign their prototypes after reviewing the collected data and the designs of the other teams.

 HELPFUL TIPS

- After the Test Trial, have teams take a gallery walk to view other teams' designs for possible ideas to assist them in the Analyze and Redesign portions of the engineering design process.

- If teams are successful on the first try, encourage them to make their prototypes even more efficient. If it is a scenario in which this is not feasible, distribute team members to other teams to be a support for them in making their prototypes more efficient. Alternatively, at teacher discretion, move students on to the Justification portion of the lesson.

- If after the third test the final prototype is still unsuccessful, have students write how they would start over. These challenges are meant to have students build on what they originally designed. If the design proved to be unsuccessful, encourage a reflection or justification on what they would do if they were allowed to start again from scratch.

REFLECTIONS — EXPLAIN & ELABORATE

AFTER TEST TRIAL 1	Which team recorded the lowest decibel level?
ANALYSIS	How can you improve your prototype design?
AFTER TEST TRIAL 2	Did the decibel level change? If so, what changes that you made to your prototype contributed to this change?
ANALYSIS	Review your data and determine if it is within the acceptable range for a restaurant. Then explain whether or not you need to make any other changes to your design.
AFTER TEST TRIAL 3	Review your data, compare it to the other teams' data, and determine if it is within the acceptable range for a restaurant.

JUSTIFICATION — EVALUATE

TECHNOLOGY — Research ways that light and sound can be programmed with computers to create a unique atmosphere for a room. Use this information to create a design for a technologically enhanced room.

ARTS — Design a mural for one side of the wall with a theme that would energize people to dance. Design a mural on the other side that would encourage tranquility. Write an explanation of how your designs meet these goals.

MATH — Use the decibel recordings gathered for all teams through the test trials to create an appropriate graphical display. Choose a histogram, line plot, or scatter plot.

LAND YACHTS AHOY!

STEAM

SETTING THE STAGE

DESIGN CHALLENGE PURPOSE

Develop a model to quickly and safely move items back and forth between two houses.

TEACHER DEVELOPMENT

To prepare students for this challenge, review Newton's three laws of motion. **Land yachts** are wind-powered wheeled vehicles that use sails to catch the wind so that they can move on land. They are enjoyed both professionally and recreationally. Land yachts are popular in the United Kingdom and Australia.

Forces have two categories, contact and noncontact. Students will have misconceptions about the difference between these types of forces, especially concerning wind. Wind physically interferes with the sail and provides the energy needed to propel a land yacht.

Therefore, wind is a contact force. **Air resistance** and **friction** are both contact forces in this lesson and can be explained using Newton's laws of motion. Students will need to understand that wind is a contact force. Noncontact forces are forces that act at a distance and do not come into physical contact with the objects that they are influencing. Review the three types of noncontact forces: electrical, gravitational, and magnetic.

Note: Visit the website listed on the inside front cover for more information about land yachts and forces.

STUDENT DEVELOPMENT

Students understand that objects move in predictable ways. However, they are now learning more about how different forces act upon these objects. It will help for students to review Isaac Newton's three laws of motion.

It will also help to discuss when they have observed objects in motion from a distance (e.g., watching playground equipment move) and when they have been on objects that move in a circle (e.g., a merry-go-round).

STANDARDS

SCIENCE	TECHNOLOGY	ENGINEERING	ARTS	MATH	ELA
MS-PS2-1	ISTE.3	MS-ETS1-1	Creating #1	CCSS.MATH. CONTENT.6.SP.B.5	CCSS.ELA-LITERACY. WHST.6-8.1
MS-PS2-2		MS-ETS1-2		CCSS.MATH. PRACTICE.MP1	
MS-PS3-1		MS-ETS1-3		CCSS.MATH. PRACTICE.MP2	
MS-PS3-2		MS-ETS1-4			

SCIENCE & ENGINEERING PRACTICES

Developing and Using Models: Develop a model to describe unobservable mechanisms.

Planning and Carrying Out Investigations: Plan an investigation individually and collaboratively, and in the design: identify independent and dependent variables and controls, what tools are needed to do the gathering, how measurements will be recorded, and how many data are needed to support a claim.

Analyzing and Interpreting Data: Construct, analyze, and/or interpret graphical displays of data and/or large data sets to identify linear and nonlinear relationships.

Constructing Explanations and Designing Solutions: Apply scientific ideas or principles to design, construct, and/or test a design of an object, tool, process, or system

CROSSCUTTING CONCEPTS

Cause and Effect: Cause-and-effect relationships may be used to predict phenomena in natural or designed systems.

Systems and System Models: Models can be used to represent systems and their interactions—such as inputs, processes, and outputs—and energy and matter flows within systems.

Stability and Change: Explanations of stability and change in natural or designed systems can be constructed by examining the changes over time and processes at different scales, including the atomic scale.

Scale, Proportion, and Quantity: Proportional relationships (e.g., speed as the ratio of distance traveled to time taken) among different types of quantities provide information about the magnitude of properties and processes.

TARGET VOCABULARY

acceleration

air resistance

distance

force (contact and noncontact)

friction

motion

speed

velocity

MATERIALS

For Land Yacht Construction:
- 1" Styrofoam ball
- toothpicks
- aluminum foil
- waxed paper
- coffee stirrers or small dowels
- glue

For Testing:
- 1 large box fan
- stopwatch
- masking tape
- cargo items to transport (6 coins, 6 cotton balls, 6 metal washers)
- tape measure
- compass
- recording sheet (page 141)

LITERACY CONNECTIONS

Give It a Push! Give It a Pull!: A Look at Forces
by Jennifer Boothroyd

NOTES

STEAM —IN— ACTION

DILEMMA　ENGAGE

Your hometown could be hit by a major storm system. The roads could become flooded, and the water will likely rise. Your home has only one level and may be covered in water. Your neighbor's house has a second level that will remain safe from the flooding. The neighbors have invited you and your family to come stay at their house in case of an emergency. You may need to grab your valuables and send them over quickly before the water reaches your house and yard. You may not have a way to move all your valuables in a timely manner without leaving things behind. Can you design a vehicle to move your valuables that travels on land and uses wind for power in case of an emergency?

MISSION

Design and construct a wind-powered land yacht to quickly and safely move items. Your prototype must travel at least 15 feet using only wind power.

BLUEPRINT　EXPLORE

Provide the Individual and Group Blueprint Design Sheets to engineering teams. Have individual students sketch a prototype to present to the other members of their team. Teams will discuss the pros and cons of each sketch and then select one prototype to construct.

ENGINEERING TASK	**TEST TRIAL**	**ANALYZE**	**REDESIGN**
Teams will construct their land yachts. *Note:* Use masking tape to mark 10 meters, 20 meters, and 30 meters from the starting point (box fan).	Teams will each test their land yacht prototypes by placing the cargo items (represents the items being moved) on the yacht and then placing the yacht at the starting point in front of the fan. Teams will use a stopwatch to determine the time it takes for their prototypes to travel the three marked distances. Teams will use the time and distance recorded to determine their prototypes' speed.	Teams will discuss the results of each distance test and compare their results with those of the other teams. Teams should be allowed to observe the other designs to gather ideas and reflect in order to improve their prototypes. *Note:* Encourage teams to think about the forces acting on their prototypes.	After analyzing their data and test results, teams will make adjustments to their prototypes. *Note:* Encourage teams to keep the forces acting on their prototypes in mind when making improvements to the design.

HELPFUL TIPS

- After the Test Trial, have teams take a gallery walk to view other teams' designs for possible ideas to assist them in the Analyze and Redesign portions of the engineering design process.

- If teams are successful on the first try, encourage them to make their prototypes even more efficient. If it is a scenario in which this is not feasible, distribute team members to other teams to be a support for them in making their prototypes more efficient. Alternatively, at teacher discretion, move students on to the Justification portion of the lesson.

- If after the third test the final prototype is still unsuccessful, have students write how they would start over. These challenges are meant to have students build on what they originally designed. If the design proved to be unsuccessful, encourage a reflection or justification on what they would do if they were allowed to start again from scratch.

REFLECTIONS — EXPLAIN & ELABORATE

AFTER TEST TRIAL 1	Which team's prototype had the best time trial? Did your land yacht safely transport the cargo items?
ANALYSIS	What was the difference in travel time between the fastest and slowest prototypes? Is there a specific feature of the fastest prototype that you think contributed to its speed? What forces caused some teams' designs to perform more slowly? What feature of the slowest team's design caused drag?
AFTER TEST TRIAL 2	Which prototype had the best time trial? Did your land yacht safely transport the cargo items?
ANALYSIS	What was the difference in travel time between the fastest and slowest prototypes?
AFTER TEST TRIAL 3	Which team of engineers had the most effective prototype?

JUSTIFICATION — EVALUATE

TECHNOLOGY	Create a spreadsheet of all of the teams' data from the test trials. Use that data to create a data plot as a graphical representation of the results.
ELA	Write an essay to convince an audience that your prototype is best suited for quickly moving items to a neighbor's house. Use scientific reasoning to justify your conclusion.
ARTS	Decorate and display your prototype in a class "museum." Then create a poster to advertise your design.
MATH	Graph your own team's data on a scatter plot, and analyze the trends in the data. Explain your conclusions based upon the data and how it might affect a possible redesign of your prototype.

SMOOTH AS ICE

STEAM

SETTING —THE— STAGE

DESIGN CHALLENGE PURPOSE

Create smooth fudge that has small-sized crystals that are not visible to the naked eye.

TEACHER DEVELOPMENT

When **matter** changes its phase, it is dependent upon **temperature**. Rapid temperature changes cause a different **rate** for the phase change than slow temperature changes. This rate change is important in several types of matter.

Matter is made of atoms, which combine to make **molecules** that vary in size and are dependent on temperature change. One such example is in the case of the growth of geodes. Geodes are igneous rocks that cool more quickly on the outside than on the inside. If the inside of the geode cools very slowly, larger crystals are formed, resulting in beautiful mineral crystals such as amethyst. If the inside of the geode cools more rapidly, the crystals are very small, resulting in a smooth inside.

Note: Just like geodes that are left undisturbed, fudge that is not immediately stirred and allowed to cool slowly, will form small crystals.

STUDENT DEVELOPMENT

Students will need to have an understanding of phase changes and the properties of each phase. Students will need to know that temperature causes those changes. They will learn from this lesson that the rate of the temperature change matters as well. Students will need to know that atoms combine to make molecules that range in size when they are combined and that temperature can affect those changes.

Note: Visit the website on the inside front cover for more information about changes in matter and how crystals are formed.

STANDARDS

SCIENCE	TECHNOLOGY	ENGINEERING	ARTS	MATH	ELA
MS-PS1-1	ISTE.1	MS-ETS1-1	Creating #1	CCSS.MATH. PRACTICE.MP1	CCSS.ELA-LITERACY. WHST.6-8.3
	ISTE.3	MS-ETS1-2	Creating #2		CCSS.ELA-LITERACY. WHST.6-8.9
	ISTE.6	MS-ETS1-3	Creating #3		
		MS-ETS1-4			

SCIENCE & ENGINEERING PRACTICES

Developing and Using Models: Develop and/or use a model to predict and/or describe phenomena.

CROSSCUTTING CONCEPTS

Scale, Proportion, and Quantity: Time, space, and energy phenomena can be observed at various scales using models to study systems that are too large or too small.

TARGET VOCABULARY

atom

cooling rate

crystals

matter

molecule

temperature

MATERIALS

- refrigerator
- freezer
- Bunsen burner or hot plate
- heat-safe bowl
- spoon
- mixing bowl
- nonstick cooking spray
- semisweet chocolate chips
- sweetened condensed milk
- butter
- nuts (optional)
- recipe (page 142)

LITERACY CONNECTIONS

Growing Crystals by Ann O. Squire

Crystals and Crystal Growing for Children: A Guide and Introduction to the Science of Crystallography and Mineralogy for Kids by Samuel Grundy-Tenn

NOTES

STEAM
—IN—
ACTION

DILEMMA | ENGAGE

Rocky Offroad has a problem! He owns a popular fudge factory, but his fudge has been coming out lumpy, causing his sales to slump. He has no idea why the fudge has changed. The recipe has not changed in 10 years, yet the fudge is acting funny. Rocky has looked over all of his paperwork and noticed only one difference: the air conditioner was broken during one week of production, and his poor employees were working in hot conditions. This change in temperature was the only change that Rocky found. He needs your help. Can your team help Rocky make his fudge smooth as ice once again?

MISSION

Design a cooling plan for making smooth fudge.

BLUEPRINT | EXPLORE

Provide the Individual and Group Blueprint Design Sheets to engineering teams. Have individual students write a detailed cooling plan to present to the other members of their team. Teams will discuss the pros and cons of each plan and then select one plan to follow.

ENGINEERING TASK	TEST TRIAL	ANALYZE	REDESIGN
Teams will follow the fudge recipe provided and then develop their own cooling plans to create smooth fudge. *Note:* Adult supervision is required while teams heat the ingredients.	Teams will follow their cooling plans and observe the size of the crystals in their fudge. *Note:* If the crystals are large enough, students should select three of the largest ones to measure. Use a microscope, if available, to view the crystals.	Teams should compare the smoothness of their fudge (smallness of crystal size) with that of the other groups. Teams should also discuss how their cooling plans compare to other teams' plans.	Students will determine if they can improve the smoothness (reduce the crystal size) of their fudge.

HELPFUL TIPS

- After the Test Trial, have teams take a gallery walk to view other teams' designs for possible ideas to assist them in the Analyze and Redesign portions of the engineering design process.

- If teams are successful on the first try, encourage them to make their prototypes even more efficient. If it is a scenario in which this is not feasible, distribute team members to other teams to be a support for them in making their prototypes more efficient. Alternatively, at teacher discretion, move students on to the Justification portion of the lesson.

- If after the third test the final prototype is still unsuccessful, have students write how they would start over. These challenges are meant to have students build on what they originally designed. If the design proved to be unsuccessful, encourage a reflection or justification on what they would do if they were allowed to start again from scratch.

STEAM

REFLECTIONS — EXPLAIN & ELABORATE

AFTER TEST TRIAL 1	How does your team's fudge look? Does it have a smooth texture? Are there visible crystals large enough to measure?
ANALYSIS	Is there a way to make smaller crystals? What was different about the more successful and less successful fudge?
AFTER TEST TRIAL 2	Does your team's fudge look smoother than during the first trial? Can you measure the crystals? If so, are they smaller than before?
ANALYSIS	Is this batch of fudge smoother than the first batch? What did you do differently this time?
AFTER TEST TRIAL 3	Is your team's fudge smoother than during the previous trial?

JUSTIFICATION — EVALUATE

TECHNOLOGY	Create a digital presentation persuading Mr. Offroad to choose your cooling technique.
ELA	Write a letter persuading Mr. Offroad to use your cooling technique in his factory.
ARTS	Create a visual presentation to convince Mr. Offroad to select your cooling technique.

ZIPPING ALONG

1-2 HOURS

TIME FOR COMPLETION

STEAM

SETTING —THE— STAGE

DESIGN CHALLENGE PURPOSE

Design and construct a zip line that quickly carries a load across a space without damaging the load.

TEACHER DEVELOPMENT

This challenge focuses on how an object's motion depends on the sum of the forces on the object and the mass of the object. The greater the mass of an object, the greater the force need to move it. **Gravitational potential energy** is the amount of energy an object has due to its position in a gravitational field. This means the higher an object, the greater its gravitational potential energy. The greater the gravitational potential energy of an object, the more **kinetic energy** the object will have when it is in motion.

When students create their zip line, they will discover that if they make their zip line too steep, they will increase the speed and risk crashing at the other end of the zip line.

Note: Visit the website on the inside front cover for more information about gravitational potential energy and kinetic energy.

STUDENT DEVELOPMENT

In order to complete this challenge, students must have an understanding of how **gravity** acts upon an object. Students should understand that gravity pulls an object toward the center of the earth. Students must understand that the more mass an object has, the more force is needed to move that object. They will need to increase or decrease the angle of their zip line in order to accurately drop their marble on the target before it reaches the end of the line.

Lesson Idea: As preparation for this challenge, have students let go of a marble from the top of a short ramp (a textbook propped up on one end). Hold the ramp at different angles to observe changes in the marble's motion, speed, and distance traveled. Have students record the speed of the marble when the ramp is held at different heights. Then have students try releasing objects of different weights, such as a ping-pong ball and a golf ball. Discuss and compare how mass affects speed and motion.

STANDARDS

SCIENCE	TECHNOLOGY	ENGINEERING	ARTS	MATH	ELA
MS-PS2-2	ISTE.1	MS-ETS1-1	Creating #1	CCSS.MATH.CONTENT.6.EE.C.9	CCSS.ELA-LITERACY.W.6.7
	ISTE.6	MS-ETS1-2	Creating #2	CCSS.MATH.CONTENT.7.G.A.2	CCSS.ELA-LITERACY.W.7.7
		MS-ETS1-3	Creating #3	CCSS.MATH.CONTENT.8.EE.B.5	CCSS.ELA-LITERACY.W.8.7
		MS-ETS1-4		CCSS.MATH.PRACTICE.MP4	
				CCSS.MATH.PRACTICE.MP5	
				CCSS.MATH.PRACTICE.MP6	

SCIENCE & ENGINEERING PRACTICES

Planning and Carrying Out Investigations: Plan an investigation individually and collaboratively, and in the design: identify independent and dependent variables and controls, what tools are needed to do the gathering, how many measurements will be recorded, and how many data are needed to support a claim.

CROSSCUTTING CONCEPTS

Stability and Change: Explanations of stability and change in natural or designed systems can be constructed by examining the changes over time and forces at different scales, including the atomic scale.

TARGET VOCABULARY

angle

distance

force

gravitational potential energy

gravity

motion

speed

MATERIALS

- fishing line (at least 4 m per team)
- 1 paper or plastic cup (or the bottom half of a water bottle)
- 10 paper clips
- 1 hard-boiled egg (the load)
- masking tape
- clear tape
- scissors
- ruler
- tape measure
- meterstick
- stopwatch
- protractor

LITERACY CONNECTIONS

The Gripping Truth about Forces and Motion
by Agnieszka Biskup

NOTES

STEAM —IN— ACTION

DILEMMA ENGAGE

Mr. Cliff Soarin wants to take his zip line adventure park to a new level. Currently, zip lines need to be able to stop themselves prior to reaching the other end. Riders can be injured if the zip lines don't stop in time. Problems can also occur if the zip line stops too soon. When this happens, riders must physically pull themselves the remaining distance of the zip line. Mr. Soarin wants to create a zip line that anyone can safely enjoy. He needs your help to design a zip line that can carry a rider safely across a distance of at least 4 meters as quickly as possible.

MISSION

Create the fastest zip line that safely carries a load (egg) across a distance of at least 4 meters.

BLUEPRINT EXPLORE

Provide the Individual and Group Blueprint Design Sheets to engineering teams. Have individual students sketch a prototype of the zip line (including the angle measurement created between the line and the ground at the starting point) to present to the other members of their team. Teams will discuss the pros and cons of each sketch and then select one prototype to construct.

Note: Students should include specific angle measurements of 45 to 90 degrees.

ENGINEERING TASK	TEST TRIAL	ANALYZE	REDESIGN
Teams will construct their zip lines to carry a load (egg) a distance of at least 4 meters. They will set the angles of the zip lines to make the load travel as fast as possible without damaging the load or the zip lines. *Note:* Teacher will mark the starting and ending points 4 meters apart. Pose questions to teams as they construct their zip lines. Ask questions such as *How might raising or lowering the ending or starting point of the zip line impact the speed?*	Teams will attach their loads to their zip lines and test their prototypes, recording observations. They should measure and record the distance their load traveled on the zip line from the starting point and the speed it traveled. Teams should use the formula for speed $$speed = \frac{distance}{time}$$ to calculate the speed of their load by using the time it took to travel from the starting point to the place it stopped.	Teams will analyze the results of the test and make changes to improve their prototypes.	After analyzing their data and test results, teams will make adjustments to their prototypes.

 HELPFUL TIPS

- After the Test Trial, have teams take a gallery walk to view other teams' designs for possible ideas to assist them in the Analyze and Redesign portions of the engineering design process.

- If teams are successful on the first try, encourage them to make their prototypes even more efficient. If it is a scenario in which this is not feasible, distribute team members to other teams to be a support for them in making their prototypes more efficient. Alternatively, at teacher discretion, move students on to the Justification portion of the lesson.

- If after the third test the final prototype is still unsuccessful, have students write how they would start over. These challenges are meant to have students build on what they originally designed. If the design proved to be unsuccessful, encourage a reflection or justification on what they would do if they were allowed to start again from scratch.

REFLECTIONS — EXPLAIN & ELABORATE

AFTER TEST TRIAL 1	What were the results of your first test? How far did your load travel? How fast did it travel? What differences or similarities do you see in the designs of the different teams?
ANALYSIS	What changes could you make to the height of the zip line (angle of the line) that might increase the success of your prototype?
AFTER TEST TRIAL 2	What were the results of your second test? How far did your load travel? How fast did it travel? What differences or similarities do you see in the designs of the different teams?
ANALYSIS	What changes could you make to the height of the zip line (angle of the line) that might increase the success of your prototype?
AFTER TEST TRIAL 3	Which team of engineers had the most effective prototype? What were the differences between the prototypes?

JUSTIFICATION — EVALUATE

TECHNOLOGY	Create a slideshow presentation about your prototype.
ELA	Conduct a short research project about zip lines. Include information on their history, popularity, and safety.
ARTS	Create a poster that advertises your zip line. Include a name and slogan.

MATH	Write math problems that include the elements listed below. Provide the solution for each problem.
	    a. the formula for calculating speed
	    b. the use of variables in equations and inequalities (supply speed and time, but ask for distance)
	    c. a distance-time graph with questions based on the graph

APPENDIX

Lesson Plan–Specific Reproducibles 129

Individual Blueprint Design Sheet 143

Group Blueprint Design Sheet 144

Budget Planning Chart 145

STEAM Job Cards 146

STEAM Rubric 147

Bibliography 149

	OUTSTANDING 4	GOOD 3	FAIR 2	POOR 1	SELF-ASSESSMENT	PEER/TEACHER ASSESSMENT
CRITICAL THINKING	• Thoughtfully and accurately interprets results • Shows in-depth understanding of major ideas	• Identifies relevant arguments • Justifies results • Offers reasons	• Usually justifies results • Usually offers reasons	• Misinterprets data • Does not justify arguments		
QUALITY OF INFORMATION	• Thoroughly covers topics • Includes details that support the topic	• Includes essential information • Includes some supporting details	• Includes most essential information • Does not provide enough details	• Lacks essential information		
ORGANIZATION	• Well organized and coherent • Topics are in logical sequence • Includes clear introduction and conclusion	• Organized • Some topics are out of logical order • Introduction and conclusion are mostly clear	• Some organization • Topics do not follow logical order • Introduction and conclusion are unclear	• Not organized • Topics make no sense		
GRAMMAR AND SPELLING	• All grammar and spelling are correct	• 1–2 grammar and spelling errors	• 2 or more grammar and spelling errors	• Frequent grammar and spelling errors		
VISUAL DESIGN	• Visually appealing • Clean simple layout • Text is easy to read • Graphics enhance understanding of ideas	• Visually appealing • Mostly clean layout • Text is easy to read • Graphics do not distract from understanding of ideas	• Text is sometimes hard to read • Some graphics are distracting	• Text is very difficult to read • Layout is cluttered and confusing		
ORAL PRESENTATION	• Well prepared • Speaks clearly • Makes consistent eye contact with audience • Smooth delivery • Invites audience questions • Accurately answers all questions	• Prepared • Speaks clearly • Makes some eye contact with audience • Answers audience questions when prompted • Accurately answers most questions	• Mostly prepared • Speaks clearly • Makes limited eye contact with audience • Accurately answers several questions	• Unprepared • Does not speak clearly • No eye contact with audience • Cannot answer questions		
TEAMWORK	• Consistently fulfills individual role in group • Sensitive to feelings and needs of group members • Assists other team members	• Fulfills individual role in group • Respectful of team members	• Works toward group goals with occasional prompting • Maintains positive attitude	• Works toward group goals only when prompted • Needs occasional reminders to be sensitive to others		
				TOTAL		

COMMENTS

Our team really liked

_____

_____

_____

The accuracy

_____

_____

_____

We suggest

_____

_____

_____

COMMENTS

Our team really liked

_____

_____

_____

The accuracy

_____

_____

_____

We suggest

_____

_____

_____

	5 POINTS	4 POINTS	3 POINTS	2 POINTS	1 POINT	0 POINTS
INDEX FOSSILS	The statistical data shows the evidence of at least three similar index fossils on both continents. The data was obtained from three reputable sources.	The statistical data shows the evidence of at least two similar index fossils on both continents. The data was obtained from two reputable sources.	The statistical data shows the evidence of at least one similar index fossil on both continents. The data was obtained from two reputable sources.	The statistical data shows the evidence of at least one similar index fossil on both continents. The data was obtained from one reputable source.	The data is not statistical but shows the evidence of at least one similar index fossil on both continents. The data was obtained from reputable sources.	The data is not statistical and is from unverified sources.
GEOLOGIC FEATURES	The statistical data shows evidence of at least three similar rocks (geologic features) on both continents. The data was obtained from three reputable sources.	The statistical data shows evidence of at least two similar rocks (geologic features) on both continents. The data was obtained from two reputable sources.	The statistical data shows evidence of at least one similar rock (geologic feature) on both continents. The data was obtained from two reputable sources.	The statistical data shows evidence of at least one similar rock (geologic feature) on both continents. The data was obtained from one reputable source.	The data is not statistical but shows the evidence of at least one similar rock (geologic feature) on both continents. The data was obtained from reputable sources.	The data is not statistical and is from unverified sources.
DIAMONDS	The statistical data shows at least three locations of diamonds on both continents. The data was obtained from three reputable sources.	The statistical data shows at least two locations of diamonds on both continents. The data was obtained from two reputable sources.	The statistical data shows at least one location of diamonds on both continents. The data was obtained from two reputable sources.	The statistical data shows at least one location of diamonds on both continents. The data was obtained from one reputable source.	The data is not statistical but shows at least one location of diamonds on both continents. The data was obtained from reputable sources.	The data is not statistical and is from unverified sources.
SEAFLOOR STRUCTURES	The display has at least five labeled seafloor structures, including continental shelves, slopes, ridges, fracture zones, and trenches.	The display has at least four labeled seafloor structures, including continental shelves, slopes, ridges, fracture zones, and trenches.	The display has at least three labeled seafloor structures, including continental shelves, slopes, ridges, fracture zones, and trenches.	The display has at least two labeled seafloor structures, including continental shelves, slopes, ridges, fracture zones, and trenches.	The display has at least one labeled seafloor structure, including continental shelves, slopes, ridges, fracture zones, and trenches.	The display has no labeled seafloor structures.
TOTAL						

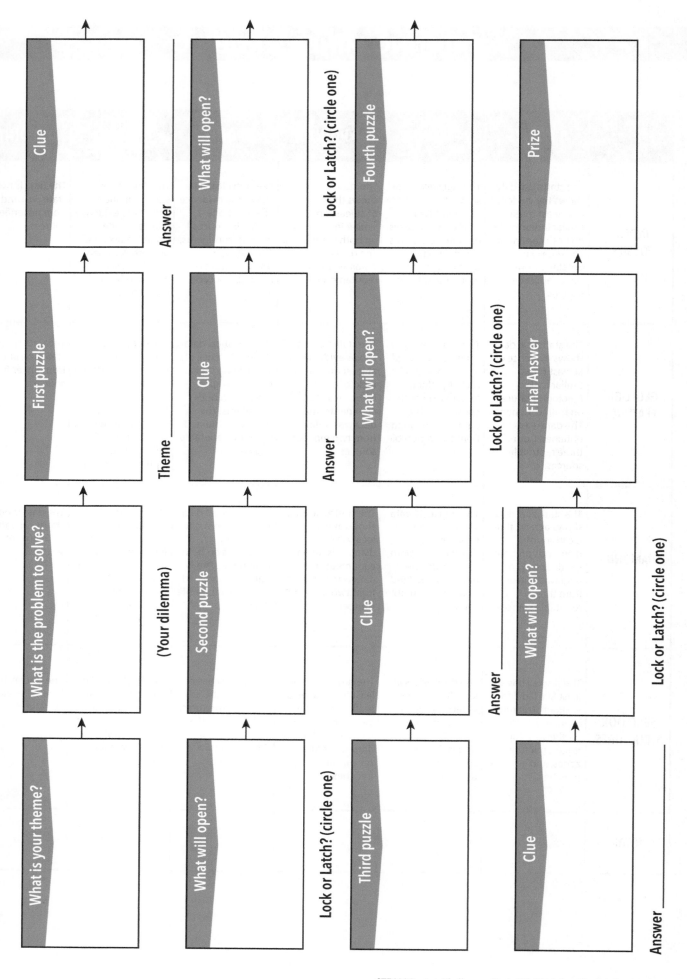

What is your theme?

What is the problem to solve?

First puzzle

Clue

(Your dilemma)

Theme _____

What will open?

Second puzzle

Clue

What will open?

Answer _____

Lock or Latch? (circle one)

Third puzzle

Clue

What will open?

Answer _____

Fourth puzzle

Lock or Latch? (circle one)

Clue

Answer _____

Lock or Latch? (circle one)

Final Answer

Prize

Lock or Latch? (circle one)

What will open?

Answer _____

Directions: Rate each component question and give constructive feedback.

	SCORE (1 = LOWEST, 3 = HIGHEST)	COMMENTS AND SUGGESTIONS
The theme of the rooms was intriguing and obvious.		
There were at least 5 problems per room that made sense and were enjoyable to solve.		
The clues and locks challenged us but were appropriate.		
If these were actual escape rooms, we would purchase a ticket.		
Newton's third law of motion was necessary to understand these rooms.		
One of the problems demonstrated electric, magnetic, or gravitational force.		
A statistical mathematical problem was included.		
The game rules were easy to understand and follow.		
TOTAL		

WE BUILT A ZOO – RUBRIC

	STRUCTURE AND WATERING HOLE REQUIREMENT	PERCENTAGE REQUIREMENT FOR THE ENCLOSURE ACREAGE	SHELTER REQUIREMENT	FEATURES OF A NATURAL HABITAT REQUIREMENT	LABELS OF STRUCTURES REQUIREMENT	LABELS OF LAND USE REQUIREMENT
3	There are at least 2 cylindrical structures and a watering hole.	A least 75% of the 2 acres was used for the enclosure.	At least 25% of the enclosure is designated as shelter.	There are at least 5 features that support a natural habitat for the animal.	All structures have labels.	Structures are labeled with the amount of land used.
2	There is 1 cylindrical structure and a watering hole.	A least 50% of the 2 acres was used for the enclosure.	At least 20% of the enclosure is designated as shelter.	There are at least 3 features that support a natural habitat for the animal.	Some structures have labels.	Some structures are labeled with the amount of land used.
1	There is only a watering hole.	A least 25% of the 2 acres was used for the enclosure.	At least 10% of the enclosure is designated as shelter.	There is at least 1 feature that supports a natural habitat for the animal.	Few structures have labels.	Few structures are labeled with the amount of land used.
TOTAL POINTS						
GRAND TOTAL						

- -

	STRUCTURE AND WATERING HOLE REQUIREMENT	PERCENTAGE REQUIREMENT FOR THE ENCLOSURE ACREAGE	SHELTER REQUIREMENT	FEATURES OF A NATURAL HABITAT REQUIREMENT	LABELS OF STRUCTURES REQUIREMENT	LABELS OF LAND USE REQUIREMENT
3	There are at least 2 cylindrical structures and a watering hole.	A least 75% of the 2 acres was used for the enclosure.	At least 25% of the enclosure is designated as shelter.	There are at least 5 features that support a natural habitat for the animal.	All structures have labels.	Structures are labeled with the amount of land used.
2	There is 1 cylindrical structure and a watering hole.	A least 50% of the 2 acres was used for the enclosure.	At least 20% of the enclosure is designated as shelter.	There are at least 3 features that support a natural habitat for the animal.	Some structures have labels.	Some structures are labeled with the amount of land used.
1	There is only a watering hole.	A least 25% of the 2 acres was used for the enclosure.	At least 10% of the enclosure is designated as shelter.	There is at least 1 feature that supports a natural habitat for the animal.	Few structures have labels.	Few structures are labeled with the amount of land used.
TOTAL POINTS						
GRAND TOTAL						

Directions: Cut along the perimeter of each grid. Then tape the two grids together to create 2 acres of land.

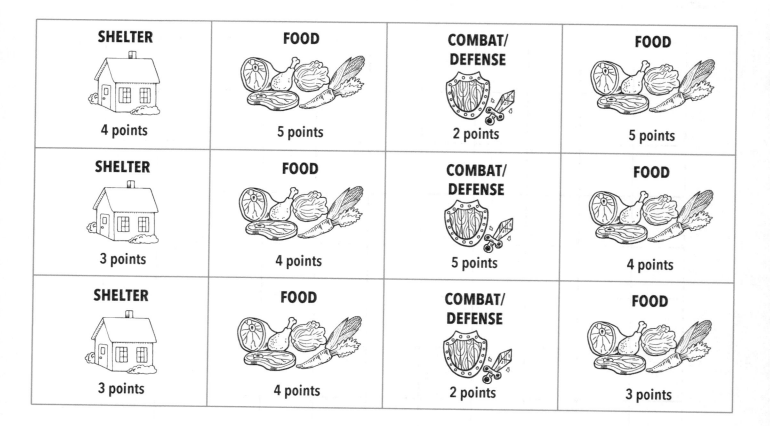

One of planet Amazon's carnivore species, the AyeNye, lives in the trees. It eats the small grubs and insects that are found inside tree branches. Its population was unaffected by the plague, and with the loss of so many predators, its population has grown significantly.

- If your animal does not live in the trees or eat insects from the trees, you are safe.

- If your animal eats the insects and/or grubs that live in trees, you will need to give up one food card to survive.

- If your animal lives in the trees, you will need to give up one shelter card to survive.

The animal populations that live in planet Amazon's rivers were the hardest hit by the epidemic, especially the smaller species, and the large carnivores and omnivores struggle to find food.

- If your animal lives on land and does not eat animals from the river, you are safe.

- If your animal lives on land but eats animals from the river then you must give up one food card to survive.

- If your animal lives in the river but is an herbivore, then you must give up one shelter card or one combat/defense card to survive.

- If your animal lives in the river and is either an omnivore or a carnivore, you must give up either one combat/defense card or one food source card.

STEAM Design Challenges Gr. 6–8 © 2018 Creative Teaching Press

One of planet Amazon's nocturnal carnivore species, the Leopardas Catan, struggles to find food as it competes for available prey.

- If your animal is a diurnal predator, you are safe.

- If your animal is a nocturnal predator, you must give up one food resource card to survive.

- If your animal is an herbivore or omnivore, you must either have a defense adaptation, such as mimicry, or give up one shelter card to survive.

Planet Amazon's largest diurnal predator species, the Caimenaz, lives and hunts for food in local rivers. These predators are ferocious and often eat other carnivores that approach the water looking for food.

- If your animal lives on land and does not hunt animals that live in rivers, you are safe.

- If your animal lives in the river, you need either a type of camouflage (or defense adaptations) or one habitat card to survive.

- If your animal eats animals that live in the river, you need to give up one combat/defense card and one food card to survive.

Planet Amazon's most common herbivore species, the Slothus, has reproduced to the point of over-crowding and overgrazing of plant life due to the extinction of its greatest predator, the Harpyeagl.

- If your animal is a carnivore or if it lives in the water, you are safe.

- If your animal is an herbivore that lives on land, you must give up one shelter card and one food card.

- If your animal is an omnivore that eats animals and/or plants that live in the water, you must give up one shelter card.

Planet Amazon's various carnivores are struggling to compete for food because so many of their usual prey died during the plague. The largest carnivore is the Jaguaran, which is over 5 feet long and 3 feet tall. It is a fierce creature with claws and sharp teeth.

- If your animal is a carnivore, you must give up one food resource card to survive and one combat/defense card unless you have wings to fly away.

- If your animal is an herbivore or omnivore, you must give up one shelter card or one combat/defense card unless you have a mimicry or camouflage adaptation that will help you to escape the Jaguaran.

Quick! Back to the drawing board! I'm starved! 1 point	So close! Try again! We're getting hungry! 2 points	Success! We eat soon! 3 points
• Equations on both the plan and the prototype don't balance. • Many of the needed components are present, but only some of them correctly model the formula for photosynthesis. • Teams failed to include factors such as the chemical formula for the atmosphere on Mars.	• Either the plan or the prototype successfully models the formula for photosynthesis, but not both. • Includes all the elements and factors, such as the chemical formula for the atmosphere on Mars.	• Both the plan and the prototype successfully model photosynthesis. • All components are clearly labeled and the process is evident. • All elements, including the chemical formula for Mars's atmosphere, are present and identifiable.
TOTAL		

- -

Quick! Back to the drawing board! I'm starved! 1 point	So close! Try again! We're getting hungry! 2 points	Success! We eat soon! 3 points
• Equations on both the plan and the prototype don't balance. • Many of the needed components are present, but only some of them correctly model the formula for photosynthesis. • Teams failed to include factors such as the chemical formula for the atmosphere on Mars.	• Either the plan or the prototype successfully models the formula for photosynthesis, but not both. • Includes all the elements and factors, such as the chemical formula for the atmosphere on Mars.	• Both the plan and the prototype successfully model photosynthesis. • All components are clearly labeled and the process is evident. • All elements, including the chemical formula for Mars's atmosphere, are present and identifiable.
TOTAL		

- -

Quick! Back to the drawing board! I'm starved! 1 point	So close! Try again! We're getting hungry! 2 points	Success! We eat soon! 3 points
• Equations on both the plan and the prototype don't balance. • Many of the needed components are present, but only some of them correctly model the formula for photosynthesis. • Teams failed to include factors such as the chemical formula for the atmosphere on Mars.	• Either the plan or the prototype successfully models the formula for photosynthesis, but not both. • Includes all the elements and factors, such as the chemical formula for the atmosphere on Mars.	• Both the plan and the prototype successfully model photosynthesis. • All components are clearly labeled and the process is evident. • All elements, including the chemical formula for Mars's atmosphere, are present and identifiable.
TOTAL		

Organelle	3	2	1
Function due to shape.	The organelle's shape was designed correctly so that it functioned correctly.	The organelle's designed shape was similar to the real organelle.	The organelle's composition was not designed correctly, and it did not function correctly.
Function due to composition.	The organelle's composition was designed so that it functioned correctly.	The organelle's designed composition was similar to the real organelle, but not exact.	The organelle's shape was not designed correctly, and it did not function correctly.
Function due to relationships among its parts.	The organelle functioned correctly due to the relationship between its parts.	The organelle functioned similarly, but not exactly, to the real organelle due to the relationship between its parts.	The organelle did not function correctly due to the relationship between its parts.
TOTAL			

LIVE ON THE SCENE - RUBRIC

	4 POINTS	3 POINTS	2 POINTS	1 POINT	TOTAL
ACCURACY OF INFORMATION	The information was accurate and could be easily understood by the audience.	The information was mostly accurate. It included a few errors but could be understood by the audience.	Only a small amount of the information presented was accurate, and the audience was left with many questions.	The information was incorrect, and the audience didn't understand what was being shared with them.	
GROUP PRESENTATION	The presentation was energetic and all group members took part in the report.	The presentation was exciting for part of the report, and most of the group members participated.	The presentation was not very energetic, and only part of the group participated in the report.	The presentation was not energetic, and only one person presented the report.	
				GRAND TOTAL	

140

STEAM Design Challenges Gr. 6–8 © 2018 Creative Teaching Press

FIRST PROTOTYPE

	TRIAL 1	TRIAL 2	TRIAL 3	AVERAGE TIME & SPEED
DISTANCE	10 meters	10 meters	10 meters	——————————
TIME				
SPEED				

	TRIAL 1	TRIAL 2	TRIAL 3	AVERAGE TIME & SPEED
DISTANCE	20 meters	20 meters	20 meters	——————————
TIME				
SPEED				

	TRIAL 1	TRIAL 2	TRIAL 3	AVERAGE TIME & SPEED
DISTANCE	30 meters	30 meters	30 meters	——————————
TIME				
SPEED				

REDESIGNED PROTOTYPE

	TRIAL 1	TRIAL 2	TRIAL 3	AVERAGE TIME & SPEED
DISTANCE	10 meters	10 meters	10 meters	——————————
TIME				
SPEED				

	TRIAL 1	TRIAL 2	TRIAL 3	AVERAGE TIME & SPEED
DISTANCE	20 meters	20 meters	20 meters	——————————
TIME				
SPEED				

	TRIAL 1	TRIAL 2	TRIAL 3	AVERAGE TIME & SPEED
DISTANCE	30 meters	30 meters	30 meters	——————————
TIME				
SPEED				

RECIPE: HOMEMADE FUDGE

3 cups semisweet chocolate chips
1 14-ounce can sweetened condensed milk
¼ cup butter
Nonstick cooking spray
1 cup chopped walnuts (optional)

Directions: Grease a baking dish with nonstick cooking spray and set aside. Place chocolate chips, sweetened condensed milk, and butter in a large heat-safe bowl. Have your teacher heat the ingredients according to your team's plan. Once the chocolate chips are melted, stir in the nuts if desired. Pour the mixture into a greased baking dish and cool according to your team's cooling plan.

RECIPE: HOMEMADE FUDGE

3 cups semisweet chocolate chips
1 14-ounce can sweetened condensed milk
¼ cup butter
Nonstick cooking spray
1 cup chopped walnuts (optional)

Directions: Grease a baking dish with nonstick cooking spray and set aside. Place chocolate chips, sweetened condensed milk, and butter in a large heat-safe bowl. Have your teacher heat the ingredients according to your team's plan. Once the chocolate chips are melted, stir in the nuts if desired. Pour the mixture into a greased baking dish and cool according to your team's cooling plan.

RECIPE: HOMEMADE FUDGE

3 cups semisweet chocolate chips
1 14-ounce can sweetened condensed milk
¼ cup butter
Nonstick cooking spray
1 cup chopped walnuts (optional)

Directions: Grease a baking dish with nonstick cooking spray and set aside. Place chocolate chips, sweetened condensed milk, and butter in a large heat-safe bowl. Have your teacher heat the ingredients according to your team's plan. Once the chocolate chips are melted, stir in the nuts if desired. Pour the mixture into a greased baking dish and cool according to your team's cooling plan.

RECIPE: HOMEMADE FUDGE

3 cups semisweet chocolate chips
1 14-ounce can sweetened condensed milk
¼ cup butter
Nonstick cooking spray
1 cup chopped walnuts (optional)

Directions: Grease a baking dish with nonstick cooking spray and set aside. Place chocolate chips, sweetened condensed milk, and butter in a large heat-safe bowl. Have your teacher heat the ingredients according to your team's plan. Once the chocolate chips are melted, stir in the nuts if desired. Pour the mixture into a greased baking dish and cool according to your team's cooling plan.

STEAM Design Challenges Gr. 6–8 © 2018 Creative Teaching Press

INDIVIDUAL BLUEPRINT DESIGN SHEET

TEAM MEMBER NAMES	PROS OF DESIGN	CONS OF DESIGN

TEAM REASONING

TEACHER APPROVAL:

BUDGET PLANNING CHART

TITLE:

MATERIALS	COST	1st TEST TRIAL		2nd TEST TRIAL		3rd TEST TRIAL	
		ITEM(S)	AMOUNT	ITEM(S)	AMOUNT	ITEM(S)	AMOUNT

TOTAL COST:

Assigning students roles, or jobs, often helps them to collaborate by giving them some guidelines to follow. As they become more practiced at problem solving, communicating, and collaborating, they will fall into these roles naturally. In the meantime, we've provided these cards, which describe each job on a student's collaborative team.

Construction Specialist

Description: This person is the one whose design was chosen. This person builds the prototype and is responsible for ensuring that the prototype follows the design parameters exactly.

Material Resource Officer

Description: This person is in charge of procuring, measuring, and cutting materials for the prototype. This person assists the construction specialist by getting materials ready and assisting in construction.

Engineering Supervisor

Description: This person is the team leader. This person assists all other team members as needed. This person acts as spokesperson for the team. This person will test the team's prototype.

Administrative Contractor

Description: This person is responsible for overseeing the construction specialist. This person must measure or otherwise ensure that prototype construction matches the blueprint design.

(Use only with groups of five.)

Accounts Manager

Description: This person holds the purse strings, keeps the team's finance records (budget sheet), and pays for all materials. This person assists the engineering supervisor with testing and recording all data.

STEAM DESIGN CHALLENGES TEAM RUBRIC

	EXEMPLARY	PROFICIENT	PROGRESSING	BEGINNING
DESIGN	Team members reach consensus as to which prototype to construct. They complete team blueprint design sheet in which they include their reasons for selecting the team prototype. They include a written explanation to compare and contrast the prototypes they sketched individually. Prototype is constructed according to specifications in the team blueprint design.	Team members reach consensus as to which prototype to construct. They include their reasons for selecting the prototype but do not include a written explanation to compare and contrast the prototypes they sketched individually. Prototype is constructed according to the specifications in the team blueprint design.	Team members reach consensus as to which prototype to construct. They include their reasons for selecting the prototype but do not include a written explanation to compare and contrast the prototypes they sketched individually. Prototype is not constructed according to the specifications of the blueprint design.	Team members reach consensus as to which prototype to construct. They do not include either their reasons for selecting the prototype or a written explanation to compare and contrast the prototypes they sketched individually. Prototype is constructed.
TEST	Team tests its prototype. Team members record observations that align with the design challenge. They make note of any unique design flaws.	Team tests its prototype and records observations that align with the design challenge.	Team tests its prototype. Team members record observations that do not align with the design challenge.	Team tests its prototype. Team members do not record observations.

STEAM DESIGN CHALLENGES TEAM RUBRIC

	EXEMPLARY	PROFICIENT	PROGRESSING	BEGINNING
ANALYZE	Team members participate in an analytic discussion about their testing and observations. They reflect on their design as compared to at least three other teams. They discuss their intended redesign steps, defending their reasoning in their discussion.	Team members participate in an analytic discussion about their testing and observations. They reflect on their design as compared to at least two other teams. They discuss their intended redesign steps.	Team members participate in an analytic discussion about their testing and observations, comparing their design with at least one other team's. They discuss their intended redesign steps.	Team members participate in an analytic discussion about their testing but do not compare their design with another team's. They discuss their intended redesign steps.
REDESIGN	Team redesigns its prototype. Original sketch is altered using a colored pencil to illustrate changes made with supporting reasons.	Team redesigns its prototype. Original sketch is altered using a colored pencil to illustrate changes made.	Team redesigns its prototype. Original sketch is altered to illustrate changes made.	Team redesigns its prototype.
EVALUATE	Team completes a justification activity. Team reflects and makes meaningful connections to the science standards as well as to two of the other STEAM standards addressed in the lesson.	Team completes a justification activity. Team reflects and makes meaningful connections to the science standards as well as to one of the other STEAM standards addressed in the lesson.	Team completes a justification activity. Team reflects and makes meaningful connections to the science standards addressed in the lesson.	Team completes a justification activity. Team makes no connection to the science standards addressed in the lesson.

BIBLIOGRAPHY

"Aeronautical Information Services - National Flight Data Center (NFDC)." Federal Aviation Administration. Accessed July 6, 2017. https://nfdc.faa.gov/xwiki/bin/view/NFDC/WebHome.

"All about Cells and Cell Structure: Parts of the Cell for Kids." YouTube video posted by FreeSchool. Accessed July 15, 2017. https://www.youtube.com/watch?v=3nBtY6LR030.

Benedictus, Leo. "Dream Jobs: Architect." The Guardian, adapted by Newsela staff. Accessed July 24, 2017. https://newsela.com/articles/dream-job-architect/id/22512/.

"Cells Cells - Parts of the Cell Rap." YouTube video posted by "CrappyTeacher." Accessed July 15, 2017. https://www.youtube.com/watch?v=-zafJKbMPA8.

"Cellular Structure and Function." PBS and WGBH Educational Foundation. Accessed July 18, 2017. https://florida.pbslearningmedia.org/resource/tdc02.sci.life.stru.lp_cell/cellular-structure-and-function/#.WW4iIOmQw2w.

Clark, Debbie. "Rube Goldberg Contraptions." Teaching Channel. Accessed July 10, 2017. https://www.teachingchannel.org/videos/rube-goldberg-contraptions.

"Earthquake Hazards Program." U.S. Geological Survey. Accessed July 8, 2017. https://earthquake.usgs.gov/.

"Ecosystems." National Geographic. Accessed July 17, 2017. https://www.nationalgeographic.org/topics/ecosystems/.

"Excerpts and Readings on Alfred Wegener." Pangaea Publishing. Accessed July 5, 2017. http://pangaea.org/wegener.htm.

"Explore Space Change Our World." The Planetary Society. Accessed July 4, 2017. http://www.planetary.org/.

"Fighter Jet Landing on Aircraft Carrier." YouTube video posted by "ciscodan123." Accessed July 8, 2017. https://youtu.be/-4UBmRNLWAc.

FlightAware flight tracker. Accessed December 18, 2016. http://flightaware.com/.

"Fun Ecosystem Facts for Kids." Easy Science for Kids. Accessed July 17, 2017. http://easyscienceforkids.com/all-about-ecosystems/.

Gonzalez, Robbie. "Explore the World's Most Detailed Map of the Seafloor, Released Today." Gizmodo Media Group. Accessed July 5, 2017. http://io9.gizmodo.com/explore-the-worlds-most-detailed-map-of-the-seafloor-r-1642315933.

Harris, Tom. "How Aircraft Carriers Work." HowStuffWorks. Accessed July 8, 2017. http://science.howstuffworks.com/aircraft-carrier4.htm.

Henderson, Sam. "How to Make a Paper Airplane." DIY Network. Accessed July 8, 2017. http://www.diynetwork.com/made-and-remade/learn-it/5-basic-paper-airplanes.

"History of the Roller Coaster." National Roller Coaster Museum and Archives. Accessed July 16, 2017. http://www.rollercoastermuseum.org/history-of-the-roller-coaster.

"How Aircraft Carriers Work." HowStuffWorks. Accessed July 8, 2017. http://science.howstuffworks.com/aircraft-carrier4.htm.

"How Can Mining Become More Environmentally Sustainable?" Fraser Institute. Accessed July 10, 2017. http://www.miningfacts.org/Environment/How-can-mining-become-more-environmentally-sustainable/.

"How Do We Measure Sound Waves?" Dangerous Decibels. Accessed July 5, 2017. http://www.dangerousdecibels.org/virtualexhibit/6measuringsound.html.

"How does a glider fly without an engine?" Discovery Communications. Accessed July 6, 2017. http://discoverykids.com/articles/how-does-a-glider-fly-without-an-engine/.

"How Old Is the Atlantic Ocean?" McDougal Littell. Accessed July 5, 2017. https://www.classzone.com/books/earth_science/terc/content/investigations/es0802/es0802page06.cfm?chapter_no=investigation.

"Human Impacts on the Biosphere." Del Mar College. Accessed July 5, 2017. http://dmc122011.delmar.edu/nsci/biology/faculty/brower/powerLectures/ch49/chapter49.pdf.

Information and Telecommunication Systems, Ohio University. Accessed July 5, 2017. www.its.ohiou.edu/mediamessage/OralPresentationRubric.doc.

"Jet-Stream of Talk." New York Times Magazine, March 7, 1954. Accessed February 10, 2017. https://groups.google.com/forum/#!topic/alt.recovery.aa/1SeDbxaUew4.

"Landings and Missed Traps F14, F18, A6a." YouTube video posted by "goldenzah." Accessed July 8, 2017. https://youtu.be/Jr9FtDi7GWU.

Loader, Pete. "Volcanoes, Molten Magma, … and a Nice Cup of Tea!" The Geological Society of London. Accessed July 5, 2017. https://www.geolsoc.org.uk/~/media/shared/documents/education%20and%20careers/volcanoes%20moltenmagma.pdf?la=en.

Martin, Erica. "How Pilots Avoid Thunderstorms." Phoenix East Aviation. Accessed July 6, 2017. https://www.pea.com/blog/posts/pilots-avoid-thunderstorms/#.

Morley, Tim. The chemical equation of photosynthesis. Accessed December 11, 2016. https://msu.edu/user/morleyti/sun/Biology/photochem.html.

"MS-ESS3-3: Earth and Human Activity." Next Generation Science Standards. Accessed July 5, 2017. https://www.nextgenscience.org/sites/default/files/evidence_statement/black_white/MS-ESS3-3%20Evidence%20Statements%20June%202015%20asterisks.pdf.

"NASA." National Aeronautics and Space Administration. Accessed July 4, 2017. http://www.nasa.gov.